LAPLACE TRANSFORMS
AND THEIR APPLICATIONS
TO DIFFERENTIAL EQUATIONS

N. W. MCLACHLAN

DOVER PUBLICATIONS, INC.
GARDEN CITY, NEW YORK

Bibliographical Note

This Dover edition, first published by Dover Publications, Inc., in 2014, is an unabridged reprint of the work originally published as *Modern Operational Calculus with Applications in Technical Mathematics* by Macmillan and Company, New York, in 1948 and reprinted in a revised edition with the title *Laplace Transforms and Their Applications to Differential Equations* by Dover in 1962.

International Standard Book Number

ISBN-13: 978-0-486-78811-1
ISBN-10: 0-486-78811-3

Manufactured in the United States by LSC Communications
4500057684
www.doverpublications.com

PREFACE TO DOVER EDITION

PRIOR to its being reprinted, the text has been revised. Additions, alterations and corrections have been made where necessary to effect improvement. The definition given at (1) §1·11 is that of the p-multiplied Laplace transform, which is used throughout the book. The reason for the presence of p outside the integral sign is given in §2·182.

LONDON, *June*, 1962.   N.W.M.

D.Sc. (ENGINEERING), LONDON; PROFESSOR OF ELECTRICAL ENGINEERING, EMERITUS, UNIVERSITY OF ILLINOIS; WALKER-AMES PROFESSOR OF ELECTRICAL ENGINEERING, UNIVERSITY OF WASHINGTON (1954).

PREFACE

THIS book is intended as an introduction to *Modern Operational Calculus* based upon the Laplace transform. It is written for post-graduate engineers and technologists. Nevertheless the purely mathematical part may be useful to advanced undergraduates in mathematics. The subject has attracted much attention during the past three decades, owing largely to its relevance to Heaviside's operational method. The Laplace transform of a function as defined by (1) § 1·11, is identical with what Heaviside called its operational form. The L.T. method of solving ordinary and partial linear differential equations is of comparatively recent date. Being logical and unambiguous, it is preferable to that of Heaviside, which may now be laid to rest in all its glory—such is the march of scientific progress!

A number of theorems or rules are established in Chapter II, the proofs usually being more complete than those given originally. Theorems 6, 6a, 9, and 14* are new. In Chapters III–VI the various theorems are used (*a*) to solve ordinary and partial linear differential equations, (*b*) to evaluate difficult integrals, (*c*) to obtain mathematical relationships and expansions, (*d*) to derive L.T.S. of various functions. Inclusion of a modern treatment of periodic impulses of finite and infinitesimal duration based upon complex integration seemed desirable. Accordingly Chapter VII is devoted to this subject, and it is a résumé of work I did during 1936–8. It is not intended for those unacquainted with complex integration, but it may prove an incentive to the study of this important branch of mathematics. The reader is advised to commence at Chapter I. No mathematical text is intended to be read from cover to cover like a novel. Accordingly the sections marked * may be omitted in a first reading, and reference made to them from time to time as the necessity arises.

* This theorem, discovered in 1940, was published in the *Mathematical Gazette* **30**, 85, 1946.

Some remarks on the ever-controversial topic of rigour may be appropriate. The text is perhaps more rigorous than is usual in a work on technical mathematics, and in this respect the technical reader is asked to consider carefully the following remarks, which are based on my own experience.

Engineers set a high standard of mechanical accuracy in limit gauges and jigs used for precision work and mass production methods. By manufacturing parts accurate to 10^{-4} inch, or even less, any individual spare, of the thousands turned out, will fit immediately into the intricate machine of which it is a component. High accuracy in machining eliminates trial and error, so a perfect fit is assured *a priori*. This being so, it is reasonable to ask engineers and technologists to accept a similar situation where mathematics is concerned. For the accuracy of the engineer is analogous to the rigour of the mathematician. Moreover, in this book the validity of operations like inverting the order of integration in a repeated integral, differentiation under the integral sign, term by term differentiation and integration of infinite series, are checked as they occur. In other words, the 'mathematical limit gauge' is used to test the analysis at various stages, so that ultimately the answer is correct without reliance upon flukes. Appendices are given so that the reader will know which 'mathematical limit gauges' are needed, when and how they should be applied.

In some technical mathematics the lack of reasonable rigour introduces uncertainty in the analysis. This, the inadequate time allotted at college, and the way in which the subject has been expounded, is largely responsible for the scepticism of engineers and engineering faculties. In the specifications and working drawings of a machine, dimensions, limits, materials, and processes of manufacture must be stated unambiguously, so that any engineer in, say, the antipodes, may understand the designer's intentions exactly. Why should this principle not apply in technical mathematics?

The problems in §§ 8·1–8·6 *should be regarded as an integral part of the book.* They contain important formulae and additional theorems, the proofs of which would have taken up too much space for inclusion in the text. Bessel functions [reference 11]

have been used freely, since they occur so often in modern
applied and technical mathematics. In fact the solutions of
a large proportion of problems therein may be expressed in
terms of exponential and Bessel functions. The reader is ex-
pected to have an elementary knowledge of the latter, e.g.
Chapters I, II, and the early part of VII in reference [11].*
   Mr. H. V. Lowry kindly read the manuscript, and I am
much indebted to him for his valuable suggestions.

LONDON, *January*, 1941.   N. W. M.

THE lapse of some seven years, between writing the MS. and
its publication, is due entirely to war and immediate post-war
conditions, which have caused such an upheaval in the printing
trade. The war has emphasised, however, the necessity for
enhanced facilities for the study of Technical Mathematics.
There is no ' chair ' of Technical Mathematics at any University
in Great Britain. The time is ripe for distinct departments of
Pure, Applied, and Technical Mathematics. The Professor of
Technical Mathematics would have to be trained in both
mathematics and technical matters, while industrial experience
and the ability to impart knowledge would be essential.
   It is appropriate to refer to the discussion on Technical
Mathematics at the Mathematical Association in April 1945.†
In general, those who lecture to undergraduate and post-
graduate engineers expressed the view that in future this subject must play a much more important part in the curriculum
than it has done hitherto. The ultimate purpose is to enable
engineers and technologists to acquire a sound but broad knowledge of mathematics and its application to technical matters.
Difficult mathematical problems requiring an intensive knowledge of certain branches of the subject should be handed over
to specialists.
   It may not be out of place to quote from the proposal I made
at the above discussion, namely, ' . . . a report on " The
Teaching of Mathematics to Engineers " be drawn up by repre-

---

*Throughout the text the numbers in [   ] indicate the references
on p. 212.
   † *Mathematical Gazette*, **29**, 145, 1945.

sentatives of the Mathematical Association and members of the leading Engineering Institutions. . . . '

Prof. T. A. A. Broadbent and Mr. A. L. Meyers have read and criticised the proofs in detail. Mr. Meyers generously undertook the laborious task of checking the whole of the analysis and the problems in §§ 8·1–8·6. I have great pleasure in recording my appreciation of their help which has been invaluable.

N. W. M.

LONDON, *May*, 1947.

CONTENTS

ix

SYMBOLS

$R(\nu)$ signifies the real part of ν : μ and ν are usually unrestricted numbers, i.e. they may be real, imaginary or complex : if real they are generally non-integral : m, n are usually positive integers : r may take any integral value including zero, which it is *convenient* to regard as a positive integer.

$f(t) \Rightarrow \phi(p)$ signifies that $\phi(p)$ is the p-multiplied Laplace transform of $f(t)$, t being real $\geqslant 0$. $f(t) \rightarrow \Phi(p)$ signifies that Φ (p) is the ordinary Laplace transform of $f(t)$ as defined at (1) §2·144. Unless stated to the contrary, or the preceding sign is used, the Laplace transforms in the book are the p-multiplied type defined by (1) § 1·11. See the first footnote on p. 2.

$f(x, t)$ signifies a function of x and t, e.g. $e^{-x^2/t}$, $x^3 t^{-1/2}$.

$\phi(p ; h_1, h_2)$ signifies a function of p, h_1, h_2 (see § 1·11).

$f(t) \sim e^{-t^2}/t$ signifies that the r.h.s. is an asymptotic formula for $f(t)$ when t is large enough.

$|x|$ signifies the modulus of x, always real and positive.

$\sum\limits_{r=0}^{\infty} |f_r(t)|$ signifies that the *moduli* of all functions in the series are to be summed.

The range $t \geqslant h$ signifies all values from $t = h$ to $t \rightarrow +\infty$, $h \geqslant 0$.

$h_1 \leqslant t \leqslant h$ signifies all values in any *closed* interval of t, and *includes* the end points h_1, h. A closed interval is *finite*.

$h_1 < t < h$ signifies all values in an open interval or range of t, excluding the end points h_1, h.

$f(t)$ is continuous in $h_1 \leqslant t \leqslant h$, signifies that the function is continuous for all values of t in the closed interval (h_1, h) : the continuity in a finite interval implies that $f(t)$ is bounded.

$t \neq 0$ signifies that t may *not* have the value zero.

$p \rightarrow +0$ and $p \rightarrow -0$ signify that p approaches zero from (a) the positive side, (b) the negative side.

$\simeq$ signifies ' approximately equal to '.

$f(t) = e^t \left. \begin{cases} 0 < t < h \\ = 0 \end{cases} \right\} t < 0, t > h$ signifies that $f(t)$ is equal to e^t in the

range $t = (0, h)$ but excluding the end points ; and is zero
for all $t < 0$ and $t > h$, being undefined at $t = 0, h$.

Heavy type, **L**, **R**, **C**, **G**, signifies inductance, resistance, etc.,
of *unit length of cable*.

When t is near to or approaches some limiting value, the
notation $f(t) = \mathbf{O}(t^{\nu})$ means that $|f(t)| < Kt^{\nu}$, K being an
absolute positive constant independent of variables or para-
meters.

$f(t) = \mathbf{O}(1)$ means that the function is bounded. When
$t \to +\infty$, $(t^2 + a^2)^2 = \mathbf{O}(t^4)$; $t(e^{-t} + 1) = \mathbf{O}(t)$; $t^2/(1 + t^2) = \mathbf{O}(1)$
with bound unity. When $t \to 0$, $(t^2 + a^2) = \mathbf{O}(1)$; $\sin t = \mathbf{O}(t)$.

The symbols for the various mathematical functions in the text
are used in reference 13, where the functions are defined.
This reference work contains an extensive list of p-multiplied
Laplace transforms for functions which occur in pure, applied
and technical mathematics.

LERCH'S THEOREM IN § 1·16.

In the Laplace transform sense, if $f(0) \neq 0$ but finite, a discontinuity occurs. Nevertheless when considered in any
interval including the origin, $f(t)$ itself may be continuous,
e.g. $\cos t$, $J_0(t)$, where $f(0) = 1$. Under these conditions, in a
broad sense (a) § 1·16 is applicable, and (conventionally) $f(t)$
may be regarded as continuous in $t \geqslant 0$. Instances of this
will be found in §§ 2·240, 3·12, 5·13, (a) in C_1 and C_2 in § 9
Appendix III.

Similarly the analysis in § 7·12 is valid if $f(t)$ is finitely discontinuous, in the sense intended in 1° § 1·15, at the ends of the
finite interval (h_1, h_2).

FOREWORD

THIS is addressed to the technical reader whose mathematical training covers a more restricted field than that of the pure mathematician. The technical reader's experience is often limited to continuous functions. Herein it is necessary to consider functions which may be either finitely or infinitely discontinuous. Differentiable functions must be continuous,* whereas integrable functions can be either continuous, or finitely and (in some cases) infinitely discontinuous. Apart from discontinuities, functions must be single valued to avoid ambiguity. If $y^2 = x$, $y = \pm x^{1/2}$ (unless $x = 0$), and the positive branch is chosen usually. It must be appreciated that *infinity is not a number*, but a limit which exceeds *any* number we care to name, however large it may be. For the sake of brevity, 'infinity' (∞) is used frequently in such a way as to appear to be a number, e.g. the upper limit in an integral. The appropriate viewpoint is that the value of the integral is required as the upper limit $\to \infty$.

In enunciating a theorem, the conditions for its validity must be stated fully. Otherwise the theorem might be used in cases where it did not hold. Sometimes a theorem may hold under conditions less stringent than those given in the proof. If stated as mere formalities, the theorems in Chapter II become brief and simple. Bereft of the conditions for their validity, they are analytically incomplete and cannot be used with confidence. Omission of logical steps in analysis is just as serious a defect as absence of credits in a cash account. The answer to a problem is essential, but *its correctness is imperative,*

---

* There are, however, certain continuous functions (not contemplated herein) which are not differentiable. Throughout the text a continuous function means one which is differentiable. Functions of the type illustrated graphically in Figs. 22, 25(b) are continuous and differentiable *between* their points of finite discontinuity. They are piecewise continuous.

a condition which can be satisfied only by analysis in the appropriate detail. By perseverance in the early stages of the text, and frequent consultation of the appropriate appendices, the reader will find that the proper mental attitude is soon acquired.

LAPLACE TRANSFORMS

AND THEIR APPLICATIONS
TO DIFFERENTIAL EQUATIONS

THE LAPLACE TRANSFORM

1·11. Definition. Consider the infinite integral

$$p \int_0^\infty e^{-pt} f(t)\, dt = \phi(p), \qquad \ldots\ldots\ldots(1)$$

p being a suitable parameter, either real or complex, while $f(t)$ is a single-valued* function integrable in every positive interval of t. t is real and $\geqslant 0$. This integral, but without the external p, was introduced into mathematical analysis by Laplace about the year 1779. $\phi(p)$, the function obtained by evaluating the integral, we *define* to be the *p-multiplied Laplace Transform* of $f(t)$. If $\phi(p)$ is given, $f(t)$ is said to be its inverse or interpretation in terms of the real variable t. When the range of integration in (1) is $t = (0, +\infty)$, ϕ is a function of p alone. If the range is $t = (h_1, h_2)$, $0 \leqslant h_1 < h_2$, ϕ is a function of p, h_1 and h_2. We then write

$$p \int_{h_1}^{h_2} e^{-pt} f(t)\, dt = \phi(p\,;\, h_1, h_2), \qquad \ldots\ldots\ldots(2)$$

the r.h.s. being defined as the p-multiplied L.T. of $f(t)$ for the interval $t = (h_1, h_2)$. Integral (2) may be written in the form (1), for if

$$F(t) = f(t) \left.\begin{matrix} \\ \\ \end{matrix}\right\} \dagger \text{ when } 0 \leqslant h_1 < t < h_2,$$
$$= 0 \quad \left.\begin{matrix} \\ \end{matrix}\right. t < h_1, \, t > h_2,$$

then

$$p \int_0^\infty e^{-pt} F(t)\, dt = \phi(p\,;\, h_1, h_2). \qquad \ldots\ldots\ldots(3)$$

Integral (1) is a particular case of (2) with $h_1 = 0$ and $h_2 \to +\infty$. When p is real and $> 0,\ddagger$ (1), (2) may be interpreted

---

* The question of $f(t)$ being continuous is discussed later. Integrability of $f(t)$ in $t = (0, h)$ implies that of $e^{-pt} f(t)$, a point to be remembered in connection with enunciation of the theorems which follow, e.g. § 1·21.

† Using Heaviside's unit function (Appendix I), this may be written in the form $F(t) = f(t)\,[H(t - h_1) - H(t - h_2)]$ also. Fig. 20 illustrates the case where $f(t) = t^3$, $h_1 = 0$, $h_2 = h$.

‡ $p > 0$ implies the reality of p, but since p may be complex sometimes, this distinctive wording is used generally.

geometrically as the areas of the exponentially damped function $pf(t)$ between the limits $t = (0, +\infty)$, $t = (h_1, h_2)$, respectively.

The p-multiplied Laplace transform of a function, as defined by (1), is usually identical with what Heaviside called its *operational form,** and the latter nomenclature is used frequently. By way of variation, some writers refer to $\phi(p)$ as the *image* of the *original* function $f(t)$. The most rational terminology seems to be that where $\phi(p)$ is regarded as the p-multiplied Laplace transform or L.T. of $f(t)$. The name Laplace must be appended, since there are other systems, where Fourier, Gauss, Hankel, Hilbert, Mellin, and Stieltjes transforms exist. Herein as elsewhere [12, 13], we shall employ the symbol $\backsimeq$, namely, a heavy u lying on its side. Thus to signify that $\phi(p)$ is the p-multiplied L.T. of $f(t)$ we write

$$\left. \begin{array}{c} f(t) \backsimeq \phi(p) \\ \phi(p) \backsimeq f(t) \end{array} \right\}^\dagger, \quad \dots\dots\dots\dots(4)$$

or

the closed end of the symbol pointing to the L.T. This is a very convenient and terse notation, the symbol being formed by a single stroke of the pen.

1·12. Example. Find the L.T. of t^ν, if $\nu = u + iv$ and $R(\nu) = u > -1$.

From (1) § 1·11 $\phi(p) = p \int_0^\infty e^{-pt} t^\nu \, dt,$ $\dots\dots\dots(1)$

$$= \Gamma(1 + \nu)/p^\nu, \dots\dots\dots\dots(2)$$

or $t^\nu \backsimeq \Gamma(1 + \nu)/p^\nu,$ $\dots\dots\dots\dots(3)$

by [12, p. 75]. Integral (1) diverges at the lower limit, unless $R(\nu) > -1$, so when $R(\nu) \leqslant -1$ the function has no L.T. $\Gamma(1 + \nu) = \int_0^\infty e^{-t} t^\nu dt$ is the gamma function introduced by Euler in 1729 [11, p. 177].

---

* The Laplace transform was defined originally without the external p, and in certain cases this is expedient (see § 2·144). There are several points in favour of the p-multiplied L.T.: (a) identity with Heaviside's operational forms, (b) the dimensional equivalence of $f(t)$ and $\phi(p)$ [see § 2·182], (c) the L.T. of a constant A is itself, and not A/p. Formulae on pp. 208–211, and in the references on p. 212 are p-multiplied L.T.S., as also are those in the text, unless stated otherwise.

† This notation refers to the L.T. for the interval $t = (0, \infty)$. For the interval $t = (h_1, h_2)$, see (2) § 1·11.

1·13. Integral equation for f(t). In this case the unknown function $f(t)$ occurs *under* the integral sign. Thus if $\phi(p)$ is known, but $f(t)$ is unknown, (1) § 1·11 is an integral equation for $f(t)$. Additional examples will be found in § 2·244 ; Chapters III, IV ; 1–5, § 8·5.

If
$$p \int_0^\infty e^{-pt} f(t)\,dt = \Gamma(1+\nu)/p^\nu, \quad\ldots\ldots\ldots\ldots(1)$$

then by § 1·12, $f(t) = t^\nu$ is *a* solution, provided $R(\nu) > -1$ to ensure convergence.

1·14. Uniqueness. We now ask if the solution given in § 1·13 is unique, i.e. is it the only solution? In reference [9] Lerch showed that if $f(t)$ is continuous in $t \geqslant 0$, but in certain cases may be unbounded at $+\infty$, it is determined uniquely by $\phi(p)$. Now t^ν is continuous in $t \geqslant 0$, but unbounded at $+\infty$, if ν is real >0, or if $R(\nu) > 0$, so with this proviso it is a unique solution of (1) § 1·13. Continuity, however, imposes an unnecessary restriction, since $f(t)$ is determined uniquely by $\phi(p)$ in the case of certain finitely discontinuous *integrable* functions, and certain infinitely discontinuous integrable functions. Before considering these functions in relation to (1) § 1·11, we shall introduce some definitions.

1·15. Discontinuous functions. A function may be discontinuous in several ways :

1°. Finitely discontinuous as in Figs. 13–17, 20–22, 25, 27–29 at the positions of the thin vertical lines. These are known as ordinary or simple discontinuities. The functions illustrated in Figs. 21 *d, e*, 22, 25 *b* may be regarded as periodic *piecewise* * continuous functions. They are integrable over a finite range $(0, t)$, and expressible in Fourier series by using the established procedure.

2°. Infinitely discontinuous like t^{-1}, $t^{-1/2}$ or $\log t$ at $t = 0$, where each function has an ' infinity '.

* A ' piecewise ' continuous function is continuous in stretches, devoid of infinities, and integrable in any finite range of t. It may $\to \infty$ with t, e.g. the ' staircase ' function of Fig. 25a. A thin vertical line at a discontinuity is conventional, and is *not* part of the graph. Near a discontinuity a function is considered as t approaches from either side. In Fig. 13, $f(t) = 0$ as $t \to -0$, $f(t) = E_0$ as $t \to +0$: at $t = 0$ it is undefined.

3°. Oscillatorily discontinuous like sin $(1/t)$ at the origin. As $t \to +0$, the function oscillates with constant amplitude, but the rate of oscillation $\to \infty$, i.e. the interval between consecutive zeros $\to 0$.

So far as L.T.S. are concerned, we shall confine our attention to the type of function in 1°, and those in 2° which fall in the category illustrated below. When

$$-1 < \nu < 0, \quad \text{or} \quad -1 < R(\nu) < 0,$$

t^ν has an infinity at the origin. Nevertheless integral (1) § 1·13 converges, and t^ν is a unique solution thereof. Other examples are $\log t$ and the Bessel functions $Y_0(t)$, $K_0(t)$, illustrated in Figs. 2, 3. As explained in § 1·211, convergence of (1) § 1·11 at the origin depends upon the 'order of infinity' of $f(t)$ as $t \to +0$ being less than unity. This condition is satisfied by $\log t$, $Y_0(t)$, $K_0(t)$, all of which are **O** $(\log t)$ when t is small and *positive*. Since

$$\int_0^h \log t \, dt = h(\log h - 1), \quad (h > 0), \quad \dots\dots\dots\dots(1)$$

the integrals

$$\int_0^h e^{-pt} \log t \, dt, \quad \int_0^h e^{-pt} Y_0(t) dt, \quad \int_0^h e^{-pt} K_0(t) dt, \quad (p > 0), \quad \dots(2)$$

converge by comparison.

Most of the functions considered herein exist when $t < 0$, but from the L.T. viewpoint, we consider the range $t \geqslant 0$ only. If $f(t)$ is the function, then for L.T. purposes we define as follows :

$$F(t) = f(t) H(t), \quad \dots\dots\dots\dots\dots\dots\dots(3)$$

where $H(t)$ is Heaviside's unit or step function treated in Appendix I. This definition is equivalent to

$$F(t) = f(t) \left.\begin{array}{l} \end{array}\right\} t > 0 \\ = 0 \quad \left.\begin{array}{l} \end{array}\right\} t < 0. \quad \dots\dots\dots\dots\dots\dots(4)$$

For convenience, however, we shall usually take $f(t)$ to signify $F(t)$ as so defined. Since $f(t) = 0$ when $t < 0$, if $f(0) = 0$, there is no discontinuity at the origin, e.g. Fig. 20. But if $f(0) \neq 0$, e.g. $\cos t$, $J_0(t)$, which have the value unity at $t = 0$, a finite discontinuity occurs (see Fig. 2 for $J_0(t)$).

*1·16. Lerch's theorem. For the benefit of those acquainted with the complex variable, we enunciate the theorem for (a) continuous, (b) discontinuous functions [9].

(a) If $f_1(t) \rightleftharpoons \phi_1(p)$ and $f_2(t) \rightleftharpoons \phi_2(p), f_1(t), f_2(t)$ being continuous in $t \geqslant 0$,* also if $\phi_1(p) = \phi_2(p)$ at all points of the half p-plane where ϕ_1, ϕ_2 are analytic (regular or holomorphic),† then $f_1(t) \equiv f_2(t)$ in $t \geqslant 0$. This means that a continuous function is determined uniquely by its L.T.

(b) If $f_1(t)$ and $f_2(t)$ are integrable functions in $t > 0$, but have discontinuities,‡ and if $\phi_1(p) = \phi_2(p)$ at all points of the half p-plane where ϕ_1, ϕ_2 are analytic (regular or holomorphic), then for $t > 0$, $f_1(t) \equiv f_2(t) + g(t)$, where $g(t)$ is a null function.

By definition $\displaystyle\int_0^t g(t)\, dt = 0$, so we may write

$$\int_0^t f_1(t)\, dt \equiv \int_0^t f_2(t)\, dt,$$

in place of $f_1 \equiv f_2 + g$. Null functions do not occur in technical applications, so they need not be considered here.

1·17. Boundedness and Continuity. The continuous function of Fig. 1a is defined by

$$\begin{aligned} f(t) &= \sin t \\ &= 0 \end{aligned} \left.\begin{aligned} & 0 \leqslant t \leqslant 2\pi, \\ & \text{for all other } t. \end{aligned}\right\}$$

$f(t)$ has an upper bound $+1$ at $t = \frac{1}{2}\pi$, and a lower bound -1 at $t = \frac{3}{2}\pi$. Moreover, it is bounded in the closed interval $0 \leqslant t \leqslant 2\pi$, and also in the infinite range $t \geqslant 0$. If M is a real constant $\geqslant 1$, then in the range $t \geqslant 0$, $M \geqslant |f(t)|$. The finitely discontinuous function of Fig. 20 has upper and lower bounds h^3 and zero, respectively, and in $t \geqslant 0$ $M \geqslant f(t)$ if $M \geqslant h^3$. Accordingly we define a bounded function in the following way : $f(t)$ is bounded in $0 \leqslant t \leqslant h$ or in $t \geqslant 0$, as the case may be, if there is a positive constant M such that $M \geqslant |f(t)|$. A bounded function

---

* In certain cases $f_1(t)$, $f_2(t)$ may be unbounded as $t \to +\infty$.

† See Fig. 4a and 1°, Appendix IV, condition (a).

‡ Since we have assumed the existence of ϕ_1 and ϕ_2, the type of discontinuity need not be defined.

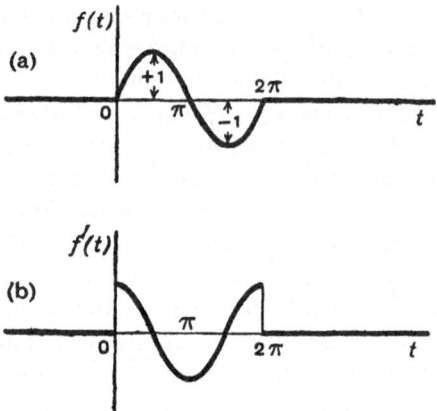

FIG. 1 (a).—Graph of the bounded continuous function

$$f(t) = \sin t \left. \right\} \; 0 \leqslant t \leqslant 2\pi$$
$$= 0 \quad \left. \right\} \text{ for all other } t.$$

The function is zero in $2\pi < t < +\infty$, and in $-\infty < t < 0$.

(b).—Graph of the finitely discontinuous function obtained by differentiating (a), namely,

$$f'(t) = \cos t \left. \right\} \; +0 \leqslant t < 2\pi$$
$$= 0 \quad \left. \right\} \; t < 0, \; t > 2\pi.$$

is not necessarily continuous in a given closed interval. For instance in Fig. 20, $F(t)$ is continuous in $0 \leqslant t \leqslant h_1$, $h_1 < h$, but discontinuous in $0 \leqslant t \leqslant h_1$, $h_1 > h$. If, however, $f(t)$ is continuous in a closed interval, it is bounded therein, but is not necessarily bounded in $t \geqslant 0$, e.g. t, t^3, e^t. As illustrated in Figs. 1, 21, a continuous function may have a discontinuous derivative. The derivative can be considered as t *approaches* the discontinuity from either side, e.g. in Fig. 20 as $t \to (h - 0)$, $F'(t) \to 3h^2$, while as $t \to (h + 0)$, $F'(t) \to 0$; but *at* $t = h$ there is no derivative.

The functions depicted in Fig. 21 a, b, c, are bounded. (a), (c) have upper bound a, lower bound zero, while those of (b) are $\pm a$. $J_0(t)$ in Fig. 2 is bounded, as also are the periodic piecewise continuous functions in Figs. 22, 25b. These six functions are all $\mathbf{O}(1)$ in $t \geqslant 0$. The finitely discontinuous function of Fig. 25a is unbounded at infinity. The functions $t^{-1/2}$, log t,

$Y_0(t)$, $K_0(t)$—see Figs. 2, 3—have infinities at the origin, whilst tan t has them at $t = (2n + 1)\pi/2$.

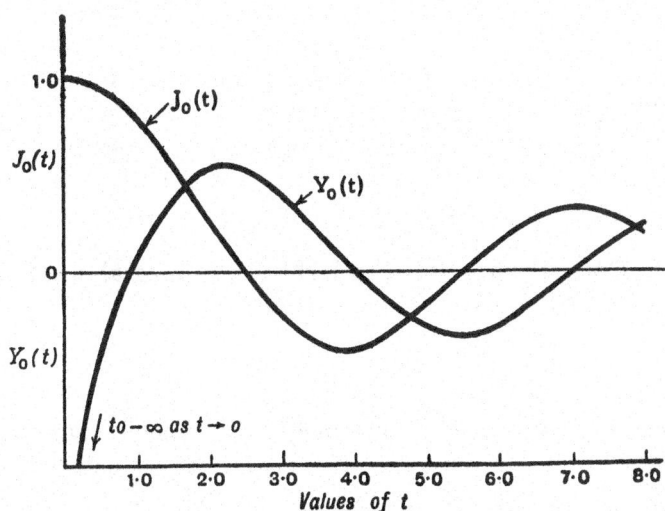

FIG. 2.—Graphs of the Bessel functions of the first and second kinds of order zero, $J_0(t)$, $Y_0(t)$. When t is small and positive, $J_0(t) \simeq 1 - \frac{1}{4}t^2$;

$Y_0(t) \simeq \frac{2}{\pi}(\gamma + \log \frac{1}{2}t)$; when $t \to \pm 0$, $J_0(t) \to 1$, $Y_0(t) = O$ (log t) and $\to -\infty$.

$\gamma = 0.5772\ldots = [1 + \frac{1}{2} + \frac{1}{3} + \ldots + \frac{1}{n} - \log_e n]_{n \to +\infty}$ is Euler's constant. When t is large enough

$$J_0(t) \sim (2/\pi t)^{1/2} \cos (t - \tfrac{1}{4}\pi), \quad Y_0(t) \sim (2/\pi t)^{1/2} \sin (t - \tfrac{1}{4}\pi).$$

$J_0(t)$ is alternating, bounded and continuous in $t \geqslant 0$. $Y_0(t)$ is alternating, bounded, and continuous in $t > t_1 > 0$; it is discontinuous and unbounded at $t = 0$, where it is undefined.

1·18. Lemmas.

1°. If $e^{-p_0 t}f(t)$ is bounded in $t \geqslant 0$, then $e^{-pt}f(t) \to 0$ as $t \to +\infty$, p real $> p_0$.

Proof. $|e^{-pt}f(t)| = |e^{-(p-p_0)t}[e^{-p_0 t}f(t)]| \leqslant Me^{-(p-p_0)t}$, ...(1)

where $M \geqslant |e^{-p_0 t}f(t)|$. Since M is constant, the r.h.s. of (1) $\to 0$ as $t \to +\infty$, if p real $> p_0$.

2°. If $e^{-p_0 t}f(t)$ is bounded in $t \geqslant 0$, then $e^{-pt}t^\nu f(t) \to 0$ as $t \to +\infty$, ν real, finite, > 0; p real $> p_0$.

Proof. $|e^{-pt}t^\nu f(t)| = |e^{-(p-p_0)t}t^\nu . [e^{-p_0 t}f(t)]|$

$$\leqslant Me^{-(p-p_0)t}t^\nu \to 0 \text{ as } t \to +\infty, \quad \ldots\ldots\ldots(2)$$

since e^x tends to infinity faster than any power of $x > 0$.

Examples.

1°. $\sin t$, $\cos t$, $J_0(t)$, are bounded in $t \geqslant 0$, so $e^{-p_0 t}f(t)$ is bounded in this range if $p_0 = 0$. Hence $e^{-pt}f(t) \rightarrow 0$ as $t \rightarrow + \infty$ if $p > 0$.

2°. The Bessel function $I_0(t)$ of Fig. 3 is bounded in $0 \leqslant t \leqslant h$, so $e^{-p_0 t}I_0(t)$ is bounded in this interval. For t large enough, $I_0(t) \sim e^t/\sqrt{2\pi t}$, so in the infinite range $t \geqslant h$, $e^{-p_0 t}I_0(t)$ is bounded if $p_0 = 1$. Hence $e^{-pt}I_0(t) \rightarrow 0$ as $t \rightarrow + \infty$ if $p > 1$. Owing to the factor $t^{-1/2}$, this also holds if $p = 1$.

1·21. Convergence of (1) § 1·11.

This integral must converge, which implies the existence of

$$\lim \int_{h_1}^{h_2} e^{-pt}f(t)\, dt, \quad h_1 \rightarrow +0, \quad h_2 \rightarrow +\infty.$$

Since e^{-pt}, p real and > 0, is a monotonic * decreasing function in $t > 0$, for convergence of (1) § 1·11 it is sufficient *but not essential* that $\int_0^\infty f(t)\, dt$ should converge. This is illustrated in (8) § 8·2 where $p = 0$. In certain cases the integral diverges although (1) § 1·11 converges, e.g. $f(t) = t^\nu$, $R(\nu) > -1$; $f(t) = e^{at}$, p real $> a > 0$. Moreover the above condition is usually too narrow, so we take a broader viewpoint.

Theorem. If $e^{-p_0 t}f(t)$ is bounded in $t \geqslant 0$,† then $\int_0^\infty e^{-pt}f(t)\, dt$ is *absolutely and uniformly convergent* with respect to p in the interval $p_1 \leqslant p \leqslant p_2$, p_1 real $> p_0$, or in the range $p > p_0$.

Proof. By hypothesis, $|e^{-p_0 t}f(t)| \leqslant M$, independent of t. If $p \geqslant p_1 > p_0$, then

$$|e^{-p_1 t}f(t)| = |e^{-p_0 t}f(t)|\,|e^{-(p_1 - p_0)t}|$$

$$\leqslant M e^{-(p_1 - p_0)t}, \quad \dotfill (1)$$

independent of p. Since $p_1 > p_0$, it follows that

$$\int_0^\infty M e^{-(p_1 - p_0)t}\, dt \quad \dotfill (2)$$

---

* See footnote to 1° § 4, Appendix III, for definition : see also Fig. 4.

† See footnote on p. 1 regarding integrability.

converges. Hence by the ' M ' test in 1° § 7, Appendix III, $\int_0^\infty e^{-pt} f(t)\, dt$ has the properties stated above. If p is complex, the theorem holds provided $R(p) > R(p_0)$.

$e^{-p_0 t} f(t)$ may have ordinary discontinuities as in § 1·17. An examination of the functions in the list or in reference 13, shows that when convergent, the convergence is *usually absolute and uniform* for $p > p_0 > 0$. It should be noted that the converse of the theorem is not necessarily true, as will be seen later in § 1·211 examples 1°, 2°, 3°, where the functions are unbounded as $t \to +0$.

★1·211. Extension of Theorem in § 1·21.

There is a class of function, unbounded at the origin, for which the integral (1) § 1·11 is *absolutely and uniformly* convergent. If $-1 < \nu < 0$, the function t^ν has an infinity at $t = 0$, but $\int_0^\infty e^{-pt} t^\nu \, dt = \Gamma(1+\nu)/p^{\nu+1}$. We can consider the convergence with reference to the ' order of the infinity ' (O.I.) at the origin. For $\mu > 0$, as $x \to +\infty$ the O.I. of x^μ is μ. Let $x = 1/t$, then as $t \to +0$, $x \to +\infty$ and the O.I. of t^ν is $-\nu$. Now for convergence, $\nu > -1$ or $-\nu < 1$, so the O.I. must be less than unity. Consequently the scope of the theorem may be extended as follows : If $e^{-p_0 t} f(t)$ is bounded in $t \geqslant 0$, or is bounded in $t \geqslant t_1$, any $t_1 > 0$, and has an infinity of order < 1 at the origin, then $\int_0^\infty e^{-pt} f(t)\, dt$ is absolutely and uniformly convergent, $p > p_0 > 0$. In some cases it may be more convenient to specify the O.I. of $f'(t)$. Since $dt^{-1}/dt = -t^{-2}$, the O.I. must be < 2.

Examples.

1°. The Bessel function $Y_0(t)$ is bounded and continuous in $t \geqslant t_1, t_1 > 0$ (see Fig. 2). When t is small, $Y_0(t) \simeq \dfrac{2}{\pi} \log t = O\,(\log t)$, and therefore $\to -\infty$ as $t \to +0$. Now $d\,(\log t)/dt = t^{-1}$, so its O.I. at $t = 0$ is unity, i.e. $\log t$ does not approach $-\infty$ as quickly as $-1/t$.

Hence $\int_0^\infty e^{-pt} Y_0(t)\, dt$ is *absolutely and uniformly* convergent in

$p > p_0 > 0$. Since $\int_0^h \log t \, dt$ exists (see § 1·15) the above is valid for $p \geqslant 0$. The value of the $Y_0(t)$ integral is

$$- [(2/\pi)/\sqrt{p^2 + 1}] \log (p + \sqrt{p^2 + 1})$$

from the list.

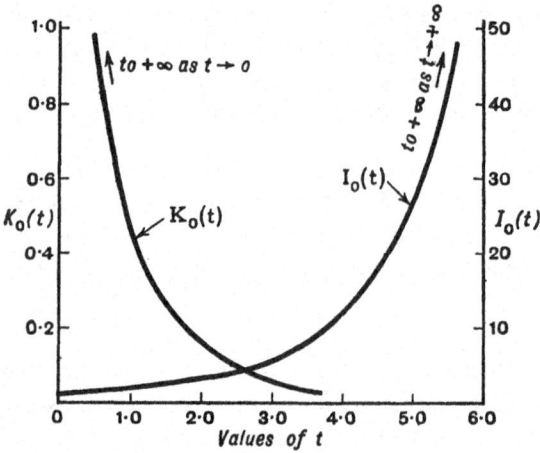

Fig. 3.—Graphs of the modified Bessel functions of the first and second kinds of zero order, $I_0(t)$, $K_0(t)$. When t is small and positive, $I_0(t) \simeq 1 + \frac{1}{4}t^2$; $K_0(t) \simeq - (\gamma + \log \frac{1}{2}t)$; when $t \to + 0$, $I_0(t) \to 1$, $K_0(t) = O [\log (t)] \to \infty$ as $t \to + 0$. When t is large enough $I_0(t) \sim e^t/(2\pi t)^{1/2}$, $K_0(t) \sim (\pi/2t)^{1/2}e^{-t}$. $I_0(t)$ is a positive monotonic increasing function, continuous in $t \geqslant 0$, unbounded as $t \to + \infty$. $K_0(t)$ is a positive monotonic decreasing function, continuous in $t \geqslant t_1$, $t_1 > 0$; it is discontinuous and unbounded at $t = 0$, where it is undefined.

2°. The Bessel function $K_0(t)$ is bounded and continuous in $t \geqslant t_1$, $t_1 > 0$ (see Fig. 3). When t is small $K_0(t) \simeq - \log t$, so by the same reasoning as above, $\int_0^\infty e^{-pt} K_0(t) \, dt$ is *absolutely and uniformly* convergent for $p \geqslant 0$. The value of the integral is

$$\frac{1}{\sqrt{p^2 - 1}} \log (p + \sqrt{p^2 - 1}).$$

3°. $e^{-p_0 t} \log t$ is bounded in $t \geqslant t_1$, $t_1 > 0$, if $p_0 > 0$, and it has an infinity of order < 1 as $t \to + 0$. Hence $\int_0^\infty e^{-pt} \log t \, dt$ is *absolutely and uniformly* convergent in $p > 0$; also in $p \geqslant 0$, from above.

4°. The Bessel function $J_\nu(t)$ of unrestricted order ν is defined for all t by

$$J_\nu(t) = \sum_{r=0}^{\infty} (-1)^r \frac{(\tfrac{1}{2}t)^{\nu+2r}}{r!\,\Gamma(\nu+r+1)}. \quad \ldots\ldots\ldots(1)$$

When ν real > -1, this series is convergent for t finite > 0, so $J_\nu(t)$ is bounded in any finite range excluding the origin. When t is large enough and $\geqslant h$

$$J_\nu(t) \sim \sqrt{2/\pi t}\,\cos\,(t - \tfrac{1}{4}\pi - \tfrac{1}{2}\nu\pi), \quad \ldots\ldots\ldots(2)$$

which is bounded in $t \geqslant h$ and $\to 0$ as $t \to +\infty$. Hence if ν real > -1, $J_\nu(t)$ is bounded in $t \geqslant t_1$, $t_1 > 0$. This is true also when ν is complex and $R(\nu) > -1$. If $R(\nu) \geqslant 0$, $J_\nu(t)$ is bounded in $t \geqslant 0$.

When $-1 < \nu < 0$, or $-1 < R(\nu) < 0$, the first term of (1), namely, $(\tfrac{1}{2}t)^\nu/\Gamma(\nu+1)$, has an infinity of order < 1, as $t \to +0$. Hence it follows from the extended version of § 1·21 that $\int_0^\infty e^{-pt}J_\nu(t)\,dt$ is absolutely and uniformly convergent in $p \geqslant 0$, if $R(\nu) > -1$. See also 2° § 8, Appendix III.

*1·212. Continuity of function represented by Laplace integral. By (1)§ 1·11 the integral below (2) § 1·21, has the value $\phi(p)/p = \Phi(p)$. Since the integral is u.c. in $p_1 \leqslant p \leqslant p_2$ or in $p > p_0$, according to § 6, Appendix III, $\Phi(p)$ is a *continuous* function of p in this range, although $f(t)$ may have ordinary discontinuities, e.g. see Figs. 20, 22, 25b. For the Morse dot function depicted in Fig. 22,

$$\int_0^\infty e^{-pt}f(t)\,dt = a/p(1 + e^{-ph}), \quad \ldots\ldots\ldots(1)$$

p real > 0. Now $f(t)$ is a periodic piecewise continuous function, but as illustrated in Fig. 4 the r.h.s. of (1) is monotonic and continuous in $p > 0$. The Fourier expansion of $f(t)$ is given at (7) § 7·15, and comprises an infinite number of continuous functions. Their sum represents $f(t)$ for all t, excepting those values where discontinuities occur, i.e. $t = nh$. According to Lerch's theorem § 1·16, the only solution of (1) as an integral equation for $f(t)$ is the Morse dot function, which is expressed analytically (except at $t=nh$) by the Fourier expansion, being integrable over the range $(0, t)$.

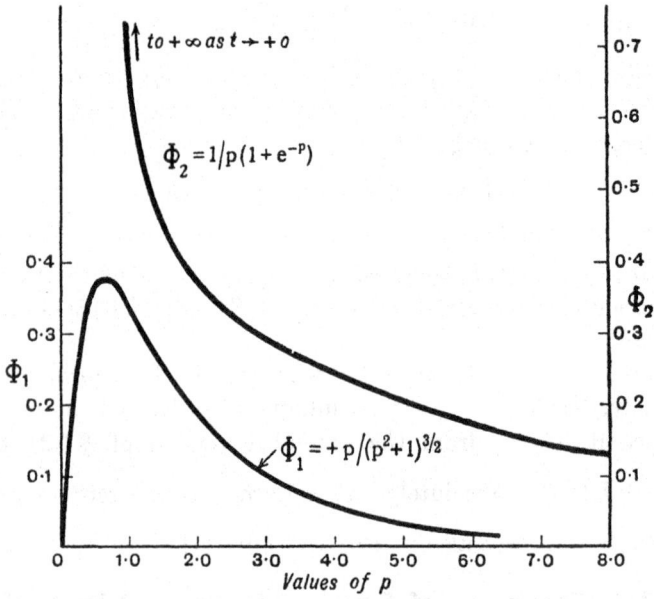

$$\Phi_2 = 1/p(1+e^{-p})$$

$$\Phi_1 = +p/(p^2+1)^{3/2}$$

Values of p

FIG. 4.—Graphs of the functions

$$\Phi_1 = p/(p^2+1)^{3/2}, \text{ and } \Phi_2 = 1/p(1+e^{-p}).$$

Φ_1 is monotonic increasing up to its maximum (upper bound) of $2/3\sqrt{3}$ at $p=1/\sqrt{2}$, and monotonic decreasing thereafter. It is positive, bounded, and continuous in $p \geqslant 0$. Φ_2 is monotonic decreasing, positive, bounded, and continuous in $p > 0$, being discontinuous and unbounded at $p = 0$.

1·213. The value of p for convergence of (I) § 1·11. If
$p = u + iv$, u and v being real, it often happens that u must be >0 for convergence. There are some cases, however, where convergence is obtained with p imaginary, i.e. $u = 0$ as in 9, 10, § 8·2. In general, if the integral converges, $u > 0$ and the value of p lies in the half plane on the right of the imaginary axis $V'OV$ (Fig. 4a). Thus we may write $p = re^{i\theta}$, $-\frac{1}{2}\pi < \theta < \frac{1}{2}\pi$, with $r = |p| = \sqrt{u^2 + v^2} > 0$. This region may be regarded as the half p-plane of convergence for the integral (if convergent). Two cases where the integral converges when $R(p) = u < 0$ are given below.

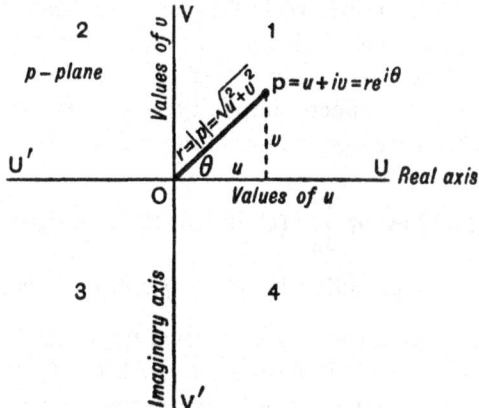

FIG. 4 (a).—Illustrating the p-plane when $p = u + iv$. The part to the *right* of the imaginary axis $V'OV$ is the half p-plane of convergence. In quadrants 1, 4, $u = R(p) > 0$; in quadrants 2, 3, $u = R(p) < 0$.

1°. Suppose p is real and $f(t) = e^{-8t}$, then if $p_0 = -8$, $e^{-p_0 t} f(t) = 1$. Thus by the theorem in § 1·21

$$\int_0^\infty e^{-pt} f(t)\, dt = \int_0^\infty e^{-(p+8)t}\, dt, \quad \ldots\ldots\ldots\ldots(1)$$

converges if $p > -8$, and it has the value $1/(p + 8)$.

2°. Let $p = u + iv$ and $f(t) = e^{-8t}$, then if

$$R(p_0) = u_0 = -8, \ |\, e^{-p_0 t} f(t)\,| = |e^{-iv_0}| = 1.$$

Hence by § 1·21 if $R(p) > -8$, the integral converges. Alternatively,

$$\int_0^\infty |\, e^{-(u+8)\,t - ivt}\,|\, dt \leqslant \int_0^\infty e^{-(u+8)t}\, dt = 1/(u + 8), \quad \ldots\ldots(2)$$

provided $u > -8$.

In both 1° and 2°, if $R(p) > -8$ it is evident that $e^{-pt} f(t) \to 0$ as $t \to +\infty$.

Generally the precise value of p need not be specified, but in certain operations it is essential that $\int_0^\infty e^{-pt} f(t)\, dt$ should converge when $p = 0$. This is treated in § 1·22. With some functions, the integral (1) § 1·11 is divergent for all p finite > 0.

Such functions do not have L.T.S. For example $t^{-1}\cos t$ is infinitely discontinuous at the origin due to the first term in the expansion, namely, t^{-1}. By § 1·12 this function has no L.T., since $\nu = -1$. To effect simplicity some of the results in Chapter II are based on p real, finite and >0.

*1·22. Evaluation of $\displaystyle\int_0^\infty f(t)\,dt$.

If this integral converges, then by 3° § 7, Appendix III, $\displaystyle\int_0^\infty e^{-pt}f(t)\,dt$ converges *uniformly* with respect to p in the closed interval $0 \leqslant p \leqslant p_1$ for every real finite value of $p_1 > 0$. In accordance with 1° § 6, Appendix III, the integral is equal to $\phi(p)/p$ for every value of p in the interval. Hence

$$\int_0^\infty f(t)\,dt = \lim_{p \to +0} \int_0^\infty e^{-pt}f(t)\,dt = \lim_{p \to +0} [\phi(p)/p], \text{ a constant. } ...(1)$$

Uniform convergence of the integral also means that $\phi(p)/p$ is continuous in the interval $0 \leqslant p \leqslant p_1$, i.e. for all values of p therein. It must not be inferred, however, that continuity of $\phi(p)/p$ implies *uniform* convergence of the integral in $0 \leqslant p \leqslant p_1$. For instance as $p \to +0$, $\phi(p)/p$ may be continuous while the integral diverges. If $f(t) = \sin t$, $\cos t$, the function alternates* with constant amplitude as $t \to +\infty$, so in accordance with §§ 1, 2, Appendix III, the integral diverges, *because it does not tend to a definite limit.*

* An alternating function is one which oscillates, having positive and negative values and, therefore, real zeros at intervals which need not be equal. $\sin t$, $\cos t$ are alternating *periodic* functions of constant amplitude. $J_0(t)$ is a damped alternating function : it does not repeat itself *exactly* at a given interval, and is, therefore, non-periodic. An oscillatory function may be (1) positive, (2) negative, (3) alternating, e.g. $(1+\cos t)$, $-(1+\sin t)$, $\sin t$. A continuous alternating or a continuous oscillatory function is represented by an alternating series, e.g. $\sin t = \displaystyle\sum_{r=0}^\infty (-1)^r \frac{t^{2r+1}}{(2r+1)!}$, $J_0(t) = \displaystyle\sum_{r=0}^\infty (-1)^r \frac{(\frac{1}{2}t)^{2r}}{(r!)^2}$.
An alternating series does not necessarily represent an alternating or even an oscillatory function, e.g. $e^{-t} = \displaystyle\sum_{r=0}^\infty (-1)^r \frac{t^r}{r!}$, this function being monotonic decreasing as t increases. Replacing $t^r/r!$ by its square, yields the series for $J_0(2t)$, which is an alternating function.

*1·23. Example. Evaluate $\int_0^{\infty} J_0(t)\,dt$.

From the list of L.T.S. on p. 209, we find that

$$\int_0^{\infty} e^{-pt} J_0(t)\,dt = 1/\sqrt{p^2 + 1}. \qquad \ldots\ldots\ldots\ldots(1)$$

In 2° § 8, Appendix III, the integral to be evaluated is shown to converge. Writing $p = 0$ in (1), it follows by § 1·22 that

$$\int_0^{\infty} J_0(t)\,dt = 1, \qquad \ldots\ldots\ldots\ldots\ldots(2)$$

which is a well-known result of importance in technical applications. The function $\phi(p)/p = 1/\sqrt{p^2 + 1}$ is monotonic * and bounded in the infinite range p real $\geqslant 0$, having the value unity when $p = 0$, and zero when $p \to +\infty$.

By aid of 2° § 8, Appendix III, the reader will easily show that $\int_0^{\infty} J_\nu(t)\,dt = 1$, provided $R(\nu) > -1$. If for simplicity we take ν real, finite, and > -1, $J_\nu(t)$ is a damped alternating function, being alternately positive and negative [see ref. 11, Figs. 2, 5].

Thus $\int_0^{\infty} J_\nu(t)\,dt$ may be interpreted as the net area between $J_\nu(t)$ and the axis of t from $t = 0$ to $+\infty$. It has the value unity for all values of ν real > -1 ; a remarkable result.

*1·231. A divergent case. When $\phi(p)/p$ is continuous in the closed interval $0 \leqslant p \leqslant p_1$ or in the infinite range $p \geqslant 0$, $[\phi(p)/p]_{p \to +0}$ has a definite value, but it must not be presumed that $\int_0^{\infty} f(t)\,dt$ is convergent. For example, if

$$f(t) = t J_0(t) \Rightarrow p^2/(p^2 + 1)^{3/2} = \phi(p), \qquad \ldots\ldots\ldots\ldots(1)$$

then $p/(p^2 + 1)^{3/2}$ is continuous and bounded in $p \geqslant 0$.† It has the value zero when $p = 0$ and $+\infty$, and attains a maximum of $2/3\sqrt{3}$ when $p = 1/\sqrt{2}$ (see Fig. 4). Nevertheless $\int_0^{\infty} t J_0(t)\,dt$ is divergent, as we shall now demonstrate.

---

* See footnote to 1° § 4, Appendix III, for definition.

† If $\Phi(p) = p/(p^2 + 1)^{3/2}$, it is a two-valued function, since

$$\Phi^2 = p^2/(p^2 + 1)^3.$$

The positive value of Φ is chosen here.

By reference [11 (33), p. 47],

$$\int_0^l t J_0(t)\, dt = \Big[t J_1(t) \Big]_0^l$$
$$= l J_1(l), \quad\quad\quad\quad\quad\quad \text{......................(2)}$$

and when l is large enough we may write [11, No.13,p.158]

$$l J_1(l) \sim \sqrt{2l/\pi}\, \cos\,(l - \tfrac{3}{4}\pi). \quad\quad\quad \text{..................(3)}$$

Thus as $l \to +\infty$, (3) does not tend to a definite limit, but oscillates over an infinite range, so the integral diverges.

Hence (1) § 1·22 cannot be applied in the case of $t J_0(t)$, since $\displaystyle\int_0^\infty e^{-pt} t J_0(t)\, dt$, $p \to +0$, is divergent. This integral is uniformly convergent in the interval $p_0 \leqslant p \leqslant p_1$, $p_0 > 0$, p_1 as large as we please. It, therefore, has the value $p/(p^2+1)^{3/2}$ in this range of p, but is not uniformly convergent as $p \to +0$, although $p/(p^2+1)^{3/2}$ is continuous in $p \geqslant 0$.

1·31. Function of $a\sqrt{t^2 - b^2}$, $t > b > 0$.

In certain technical problems involving wave motion, e.g. the theory of compressional shock waves § 4·18, of uniform transmission lines, § 4·311, and that of exponential loud-speaker horns [12], the variable has this form, b having the same dimension as t. If t represents time, then $b = x/c$, where x represents distance and c the velocity of propagation of the wave motion. The form of the variable $a\sqrt{t^2 - b^2}$ indicates that a disturbance at the origin $x = 0$, takes a time $t_1 = x/c$ to reach a point distant x therefrom. With a real and > 0, the variable $a\sqrt{t^2 - b^2}$ becomes imaginary when $t < b$. Accordingly we shall define the Laplace transform of the function $f(a\sqrt{t^2 - b^2})$ to be that of

$$\begin{matrix} F(t) = f(a\sqrt{t^2 - b^2}) \\ = 0 \end{matrix} \left.\begin{matrix} \\ \end{matrix}\right\} \begin{matrix} \text{when } t > b, \\ \text{,, } \ t < b \text{ (see Fig. 15).} \end{matrix} \quad \text{.........(1)*}$$

Thus we have

$$p \int_b^\infty e^{-pt} f(a\sqrt{t^2 - b^2})\, dt = \phi(p) \quad\quad \text{..............(2)†}$$

or
$$f(a\sqrt{t^2 - b^2}) \risingdotseq \phi(p). \quad\quad \text{..................(3)}$$

* Alternatively we may write $F(t) = f(a\sqrt{t^2 - b^2})\, H(t - b)$, by Appendix I.

† As in § 1·21 this integral is a. and u.c. in $p_1 \leqslant p \leqslant p_2$, $p_1 > p_0$ and in $p > p_0$, if $e^{-p_0 t} f(a\sqrt{t^2 - b^2})$ is bounded in $t \geqslant b$.

1·32. Evaluation of $\int_b^{\infty} f(a\sqrt{t^2 - b^2})\,dt$**.** If the integral in (2)
§ 1·31 converges uniformly as $p \to +0$, then

$$\int_b^{\infty} f(a\sqrt{t^2 - b^2})\,dt = \lim_{p \to +0}\,[\phi(p)/p], \quad \text{a constant.} \quad \text{.........(1)}$$

By 3° § 7, Appendix III, it is *sufficient* that the integral itself is convergent for (1) to hold. The remarks following (1) § 1·22 apply here also.

1·41. Operations on integral (1) § 1·11. This integral is amenable to mathematical manipulation in a variety of ways which are comparatively simple. By using the operations of partial integration, differentiation or integration under the integral sign with respect to p, and cognate operations, either separately or in combination, a number of rules or theorems can be established. Their application frequently effects a marked simplification in the solution of problems in pure, applied, and technical mathematics. This is exemplified in later chapters.

Before passing on to establish the many theorems of the (so-called) operational calculus, we refer the reader to the Appendices, where a brief résumé on convergence is given. As this subject is of vital importance herein, not only the Appendices, but also the references associated therewith should be studied. To economise space, the majority of proofs of the convergence of various integrals and series throughout the text are left to the reader as suitable exercises. By dealing with these whenever they occur, a convergence ' technique ' will soon be acquired. In fact it will be found indispensable. Inversion of the order of integration in infinite repeated integrals is also a subject of major importance, which may need practice and perseverance to master. The brief treatment in Appendix III should be supplemented by that in references [3, 4, 6, 7, 16].

II

THEOREMS OR RULES OF THE OPERATIONAL CALCULUS

2·11. Theorem I. If $f(t) = \phi(p)$,* then $f(at) = \phi(p/a)$, a real and > 0.

Proof. Since $f(t) = \phi(p)$, integral (1) § 1·11 converges for p real $> p_0$.

Writing p/a for p therein, if $p/a > p_0$, we have

$$(p/a) \int_0^\infty e^{-pt/a} f(t)\, dt = \phi(p/a). \qquad\qquad(1)$$

Let $t = \tau a$, then

$$p \int_0^\infty e^{-p\tau} f(a\tau)\, d\tau = \phi(p/a). \qquad\qquad(2)$$

Hence replacing τ by t, we obtain

$$p \int_0^\infty e^{-pt} f(at)\, dt = \phi(p/a), \qquad\qquad(3)$$

or
$$f(at) = \phi(p/a). \qquad\qquad(4)$$

2·12. Theorem 2.—The shift theorem.

If $F(t) = f(t-h)$ $\begin{cases} \text{when } t > h > 0 \\ = 0 \quad\quad t < h, \end{cases}$ then $F(t) = e^{-ph}\phi(p)$.

Proof. In (1) § 1·11 write $f(t-h)$ for $f(t)$, thereby shifting the origin to the point $t = h$. Since negative values of $(t-h)$ are excluded, the lower limit is now $t = h$. Thus (1) § 1·11 becomes

$$p \int_h^\infty e^{-pt} f(t-h)\, dt = p \int_0^\infty e^{-p(\tau+h)} f(\tau)\, d\tau, \qquad\ldots\ldots\ldots\ldots(1)$$

---

* Unless stated otherwise, this relationship and, therefore, the convergence of (1) § 1·11, and the existence of $\phi(p)$, is implied in all theorems. Also we shall sometimes tacitly assume that $\Phi(p) = \phi(p)/p$ is either differentiable or integrable n times, as the case may be, e.g. §§ 2·17, 2·21. t is always real and positive. Fairly broad conditions for convergence are given in §§ 1·21, 1·211.

with $t = \tau + h$. Replacing τ in the second integral by t, we have

$$p \int_h^\infty e^{-pt} f(t-h)\, dt = pe^{-ph} \int_0^\infty e^{-pt} f(t)\, dt = e^{-ph} \phi(p), \quad \ldots \ldots (2)$$

or $\qquad F(t) = f(t-h)\, H(t-h) \rightleftharpoons e^{-ph} \phi(p). \qquad \ldots \ldots \ldots \ldots (3)$

Graphs illustrating formula (3) will be found in Figs. 15, 16, 19, 28, 29, while $H(t)$ is treated in Appendix 1.

2·121. Example. If $\cos t \rightleftharpoons p^2/(p^2+1)$, $J_0(t) \rightleftharpoons p/\sqrt{p^2+1}$, what are the L.T.S. of $\cos(t-h)$ and $J_0(t-h)$, $0 < h < t$?

By (3) § 2·12 $\qquad \cos(t-h) \rightleftharpoons e^{-ph} p^2/(p^2+1), \qquad \ldots \ldots \ldots \ldots (1)$

and $\qquad\qquad J_0(t-h) \rightleftharpoons e^{-ph} p/\sqrt{p^2+1}. \qquad \ldots \ldots \ldots \ldots (2)$

So far as L.T.S. are concerned, these functions, presumed to be zero when $t < h$, are finitely discontinuous at $t = h$, as illustrated in Figs. 28, 29.

*2·122. An apparent anomaly. Attention is directed to the following point: Suppose the L.T. of $f(a\sqrt{t^2-b^2}) + c$, where c is a constant, is $\phi(p)$. Then we have

$$p \int_b^\infty e^{-pt} f(a\sqrt{t^2-b^2})\, dt$$

$$= p \int_b^\infty e^{-pt} [f(a\sqrt{t^2-b^2}) + c]\, dt - p \int_b^\infty e^{-pt} c\, dt, \quad \ldots \ldots (1)$$

$$= \phi(p) - ce^{-pb}. \qquad \ldots \ldots \ldots \ldots \ldots \ldots \ldots \ldots \ldots \ldots (2)$$

Thus if

$$f(a\sqrt{t^2-b^2}) + c \rightleftharpoons \phi(p), \qquad \ldots \ldots \ldots \ldots \ldots (3)$$

then $\qquad\qquad f(a\sqrt{t^2-b^2}) \rightleftharpoons \phi(p) - ce^{-pb}. \qquad \ldots \ldots \ldots \ldots (4)$

To avoid this apparent anomaly, (3) may be written in its orthodox form, namely,

$$[f(a\sqrt{t^2-b^2}) + c]\, H(t-b) \rightleftharpoons \phi(p), \qquad \ldots \ldots \ldots \ldots (5)$$

from which the relationship

$$f(a\sqrt{t^2-b^2})\, H(t-b) \rightleftharpoons \phi(p) - ce^{-pb} \qquad \ldots \ldots (6)$$

follows by § 2·12.

*2·123 Example. Formulae (3)–(6) in § 2·122 may be illustrated by aid of the modified Bessel function of the first kind of unit order. From the list we have

$$ab \int_b^t \frac{I_1(a\sqrt{t^2 - b^2})}{\sqrt{t^2 - b^2}}\, dt \coloneqq [e^{-b\sqrt{p^2 - a^2}} - e^{-bp}], \quad \dots\dots\dots(1)$$

giving

$$1 + ab \int_b^t \frac{I_1(a\sqrt{t^2 - b^2})}{\sqrt{t^2 - b^2}}\, dt \coloneqq e^{-b\sqrt{p^2 - a^2}}, \quad \dots\dots\dots(2)$$

where unity on the l.h.s. actually represents $H(t - b)\, [\coloneqq e^{-bp}]$.

2·13. Theorem 3.—The exponential multiplier theorem.

$e^{at} f(t) \coloneqq \dfrac{p}{p - a}\phi(p - a)$, where a may have any finite real or complex value.

Proof. Writing $e^{at} f(t)$ for $f(t)$ in (1) § 1·11, with $R(p - a) > p_0$,* we get

$$p \int_0^\infty e^{-(p-a)t} f(t)\, dt = \left(\frac{p}{p-a}\right) \left[(p - a) \int_0^\infty e^{-(p-a)t} f(t)\, dt \right], \quad \dots(1)$$

so

$$e^{at} f(t) \coloneqq \frac{p}{p - a}\, \phi(p - a). \quad \dots\dots\dots\dots(2)$$

2·131. Example.

Find the L.T.S. of $e^{-at} \sin \omega t$ and $e^{-at} \cos \omega t$. From the list we obtain

$$\sin \omega t \coloneqq \omega p/(p^2 + \omega^2). \quad \dots\dots\dots\dots(1)$$

Using (2) § 2·13 we get

$$e^{-at} \sin \omega t \coloneqq \left(\frac{p}{p+a}\right) \left[\frac{\omega(p+a)}{(p+a)^2 + \omega^2} \right] \quad \dots\dots\dots(2)$$

$$= \omega p/[(p+a)^2 + \omega^2]. \quad \dots\dots\dots\dots(3)$$

From the list we obtain $\cos \omega t \coloneqq p^2/(p^2 + \omega^2)$. $\quad \dots\dots\dots(4)$

Using (2) § 2·13 we get

$$e^{-at} \cos \omega t \coloneqq \left(\frac{p}{p+a}\right) \left[\frac{(p+a)^2}{(p+a)^2 + \omega^2} \right] \quad \dots\dots\dots(5)$$

$$= p(p+a)/[(p+a)^2 + \omega^2]. \quad \dots\dots\dots(6)$$

* See § 1·21.

2·14. Theorem 4.—The differentiation theorem.

If $e^{-p_0 t} f^{(r)}(t)$ is bounded and continuous in $t \geqslant 0$ for $r = 0, 1, 2, \ldots n$, then with p real $> p_0$

$$\frac{d^n}{dt^n} f(t) \doteqdot p^n \phi(p) - \sum_{r=0}^{n-1} p^{n-r} f^{(r)}(0).$$

Proof. By (1) § 1·11

$$\phi(p) = - \int_0^\infty f(t)\, d(e^{-pt}). \qquad \ldots\ldots\ldots\ldots\ldots(1)$$

Integrating by parts, from §§ 1·18, 1·21, we obtain

$$\phi(p) = - \left[e^{-pt} f(t) \right]_0^\infty + \int_0^\infty e^{-pt} f'(t)\, dt, \qquad \ldots\ldots\ldots\ldots(2)$$

$$= f(0) \qquad + \int_0^\infty e^{-pt} f'(t)\, dt, \qquad \ldots\ldots\ldots(3)*$$

if $e^{-pt} f(t) \to 0$ as $t \to +\infty$.

Rearranging (3) gives

$$p \int_0^\infty e^{-pt} f'(t)\, dt = p[\phi(p) - f(0)], \qquad \ldots\ldots\ldots\ldots\ldots(4)$$

so

$$f'(t) \doteqdot p[\phi(p) - f(0)]. \qquad \ldots\ldots\ldots\ldots\ldots\ldots(5)$$

From (5) it follows that

$$f^{(2)}(t) \doteqdot p\{p[\phi(p) - f(0)] - f'(0)\}, \qquad \ldots\ldots\ldots(6)$$

$$= p^2 \phi(p) - p^2 f(0) - p f'(0). \qquad \ldots\ldots\ldots(7)$$

Ultimately we find that

$$f^{(n)}(t) \doteqdot p^n \phi(p) - \sum_{r=0}^{n-1} p^{n-r} f^{(r)}(0). \qquad \ldots\ldots\ldots(8)\dagger$$

If $f^{(r)}(0) = 0$, for $r = 0, 1, 2, \ldots n-1$,

$$f^{(n)}(t) \doteqdot p^n \phi(p), \qquad \ldots\ldots\ldots\ldots\ldots\ldots\ldots(9)$$

and in particular

$$f'(t) \doteqdot p \phi(p). \qquad \ldots\ldots\ldots\ldots\ldots\ldots\ldots(10)$$

When the variable is $a\sqrt{t^2 - b^2}$, see 22 § 8·3.

---

* Since $e^{-p_0 t} f^{(r)}(t)$ is bounded in $t \geqslant 0$, by § 1·18 $e^{-pt} f^{(r)}(t) \to 0$ as $t \to +\infty$, and the term corresponding to the upper limit in the second member of (2) is zero.

The theorem is true if $f(t)$ is finitely discontinuous at the origin, e.g. $f(t) = H(t) \cos t$.

† For a differentiable (continuous) function of x and t, say $\theta(x\,;\,t)$,

$$\frac{\partial^n}{\partial t^n} \theta(x, t) = p^n \phi(x\,;\,p) - \sum_{r=0}^{n-1} p^{n-r} \left[\frac{\partial^r}{\partial t^r} \theta(x, t) \right]_{t=0}.$$

When $n = 1$, $\qquad \dfrac{\partial}{\partial t} \theta(x, t) = p[\phi(x, p) - \theta(x, 0)].$

2·141. Remarks relating to (5) and (10) § 2·14. Without knowing the function $f(t)$, we may desire to determine whether the L.T. of $f'(t)$ is given by (5) or by (10). If the L.T. is expansible in an absolutely convergent series in descending (negative) powers of p (see § 2·18), and the order of its denominator in p exceeds that of its numerator, then $f(0) = 0$ and (10) holds. For, each term in the expansion is of the type $p^{-\nu} \subset t^\nu / \Gamma(1 + \nu)$, $R(\nu) > 0$, so all terms vanish with t. If the first term is a constant, $f(0)$ has this value, a finite discontinuity occurs at the origin, and the L.T. of $f'(t)$ is given by (5).

★2·142. Remarks on continuity. It is usual to stipulate conditions *sufficient* for the validity of a theorem. But in certain cases one or more of the conditions may be relaxed, e.g. if $f(t) = t^{1/2}, f'(t) \Rightarrow \sqrt{\pi p}$ although $f'(t) = \frac{1}{2} t^{-1/2}$ is infinitely discontinuous at the origin. By (3) § 1·12 $\int_0^\infty e^{-pt} t^{-1/2} dt$ converges (uniformly) to the value $\sqrt{\pi / p}$ in $p_0 \leqslant p \leqslant p_1$, if p_0 real > 0. The integral corresponding to $f^{(2)}(t)$, namely, $-\frac{1}{4} \int_0^\infty e^{-pt} t^{-3/2} dt$ diverges due to the infinite discontinuity at the origin, whose order $(3/2)$ exceeds unity. Thus when $f(t) = t^{1/2}$, the theorem is invalid if $n > 1$. In general, if $f(t) = t^{-q+m}$, $0 < q < 1$, q real, m a positive integer, the integrals corresponding to $f^{(r)}(t)$ for $r = 0$, $1, 2, \ldots m$ are convergent, whereas that corresponding to $f^{(m+1)}(t)$ is divergent.★

2·143. Example. Given that

$$(a)\ \cos t \Rightarrow p^2/(p^2 + 1)\ ;\quad (b)\ J_0(t) \Rightarrow p/\sqrt{p^2 + 1},$$

find the L.T.S. of $\sin t$ and $J_1(t)$.

(a) By (5) § 2·14 $\qquad -\dfrac{d}{dt} \cos t = \sin t \Rightarrow -\dfrac{p^3}{p^2 + 1} + p, \ldots\ldots\ldots(1)$

since $\cos(0) = 1$, so $\qquad\qquad\qquad \sin t \Rightarrow p/(p^2 + 1). \ldots\ldots\ldots\ldots(2)$

By (10) § 2·14 $\qquad\qquad \dfrac{d}{dt} \sin t = \cos t \Rightarrow p \cdot p/(p^2 + 1)$

$$= p^2/(p^2 + 1), \ldots\ldots\ldots\ldots(3)$$

since $\sin(0) = 0$.

★ Reference should be made to the remarks on convergence in §§ 1·21, 1·211.

(b) By (5) § 2·14, since $J_0(0) = 1$,

$$J_1(t) = -\frac{d}{dt} J_0(t) \Rightarrow -\frac{p^2}{\sqrt{p^2+1}} + p, \quad \dots\dots\dots(4)$$

so
$$J_1(t) \Rightarrow p[\sqrt{p^2+1} - p]/\sqrt{p^2+1}, \quad \dots\dots(5)$$

$$= p/\sqrt{p^2+1}\,(p + \sqrt{p^2+1}), \quad \dots\dots(6)$$

when the numerator and denominator of (5) are each multiplied by $(\sqrt{p^2+1} + p)$.

*2·144. Modification of (8) § 2·14. The following modification is expedient in the application of this type of formula to the solution of ordinary linear differential equations with *variable* coefficients, as in Chapter VI. We shall write

$$\Phi(p) = \int_0^\infty e^{-pt} f(t)\,dt, \quad \dots\dots\dots\dots(1)$$

and
$$f(t) \rightharpoondown \Phi(p),^* \quad \dots\dots\dots\dots\dots(2)$$

where (1) is the original way of defining the Laplace transform. Then we readily find that if the conditions in § 2·14 are satisfied,

$$f^{(n)}(t) \rightharpoondown p^n \Phi(p) - \sum_{r=0}^{n-1} p^{n-r-1} f^{(r)}(0). \quad \dots\dots(3)$$

2·15. Theorem 5.—The integration theorem.

$\left[\int_0^t d\tau\right]^n f(\tau) \Rightarrow p^{-n} \phi(p)$, if the n-fold integral is convergent† and

$\lim e^{-pt} \int_0^t \left\{\left[\int_0^\tau d\tau\right]^r f(\tau)\right\} d\tau \to 0$ as $t \to +\infty$ for $r = 0, 1, 2, \dots n-1$.

Proof. Let $F_1(t) = \int_0^t f(\tau)\,d\tau \Rightarrow \phi_1(p)$, then by (5) § 2·14,

$$F_1'(t) = f(t) \Rightarrow p\,[\phi_1(p) - F_1(0)], \quad \dots\dots\dots\dots(1)$$

or
$$f(t) \Rightarrow p\,\phi_1(p), \quad \dots\dots\dots\dots\dots(2)$$

---

* The new sign $\rightharpoondown$ is used here to avoid ambiguity.

† This is certainly true if $f(t)$ is continuous or piecewise continuous in the range of integration. It is also true if $f(t) = t^\nu$, $-1 < \nu < 0$, which is unbounded at the origin (see § 2·142).

provided $\lim e^{-pt}F_1(t) = \lim e^{-pt} \int_0^t f(\tau)\,d\tau \to 0$ as $t \to +\infty$ * (see § 2·14).

Since $f(t) \rightleftharpoons \phi(p)$, it follows from (2) that

$$\phi_1(p) = \phi(p)/p \rightleftharpoons \int_0^t f(\tau)\,d\tau. \quad \dots\dots\dots\dots(3)$$

By repeated application of the above procedure, we find ultimately that if

$$F_n(t) = \left[\int_0^t d\tau\right]^n f(\tau), \quad \dots\dots\dots\dots(4)$$

then

$$F_n(t) \rightleftharpoons p^{-n}\phi(p), \quad \dots\dots\dots\dots(5)$$

provided that $\lim e^{-pt}F_1{}^{(r)}(t) = \lim e^{-pt} \int_0^t \left\{\left[\int_0^\tau d\tau\right]^r f(\tau)\right\} d\tau \to 0$ as $t \to +\infty$ for $r = 0, 1, 2, \dots n-1$. This is certainly true if $\int_0^\infty \{\ \}\,d\tau$ converges; a sufficient but not necessary condition.

*2·151. Theorem 5a.—Function of $a\sqrt{t^2-b^2}$, $t > b > 0$. If $f(a\sqrt{t^2-b^2}) \rightleftharpoons \phi(p)$, then

$$\left[\int_b^t d\tau\right]^n f(a\sqrt{\tau^2-b^2}) \rightleftharpoons p^{-n}\phi(p),$$

provided the n-fold integral exists and

$$\lim e^{-pt} \int_b^t \left\{\left[\int_b^\tau d\tau\right]^r\right\} f(a\sqrt{\tau^2-b^2})\,d\tau \to 0 \text{ as } t \to +\infty,$$

for $r = 0, 1, 2, \dots n-1$.

The proof is left as an exercise for the reader.

2·152. Example. Given that $\sin^2 t \rightleftharpoons 2/(p^2+4)$, find the L.T. of $(t - \tfrac{1}{2}\sin 2t)$.

Now $\int_0^t \sin^2 t\,dt = \tfrac{1}{2}\int_0^t (1 - \cos 2t)\,dt = \tfrac{1}{2}(t - \tfrac{1}{2}\sin 2t), \quad \dots\dots(1)$

---

* This is certainly true if $\int_0^\infty f(\tau)\,d\tau$ converges, a *sufficient* but not *necessary* condition, e.g. for $f(t) = t^n$, $e^{at}(p > a)$, $\sin t$, $\cos t$ this integral diverges, but the condition given in the proof is satisfied.

and $e^{-pt} \int_0^t \sin^2 t\, dt \to 0$ as $t \to +\infty$, so by (3) § 2·15,

$$t - \tfrac{1}{2}\sin 2t \Rightarrow 4/p\,(p^2+4). \quad\ldots\ldots\ldots\ldots(2)$$

*2·16. Theorem 6. If $\int_0^\infty f(t)\, dt$ is convergent, then

$$\int_t^\infty f(t)\, dt \Rightarrow \lim_{p\to+0}\,[\phi(p)/p] - \phi(p)/p, \quad t>0.$$

Proof. If c is any constant

$$p\int_0^\infty e^{-pt} c\, dt = c, \text{ so } c \Rightarrow c.^* \quad\ldots\ldots\ldots\ldots(1)$$

By (1) § 1·22 convergence of the integral implies that

$$\int_0^\infty f(t)\, dt = \lim_{p\to+0}\,[\phi(p)/p] = \text{a constant.}$$

Hence using (3) § 2·15, we have

$$\int_0^\infty f(t)\, dt - \int_0^t f(t)\, dt \Rightarrow \int_0^\infty f(t)\, dt - \phi(p)/p, \quad\ldots\ldots\ldots(2)$$

so

$$\int_t^\infty f(t)\, dt \Rightarrow \lim_{p\to+0}\,[\phi(p)/p] - \phi(p)/p. \quad\ldots\ldots(3)$$

If $f(t)$ takes the form $g(t)/t$, then with $g(t) = \cos t$, $J_0(t)$, the integral $\int_0^\infty g(t)\, dt/t$ diverges at the origin due to the integrand being infinitely discontinuous. However, $\int_t^\infty$ converges for $t>0$, and the L.T. can be found using § 2·23.

*2·161. Example. Find the L.T. of $\int_t^\infty J_0(t)\, dt$, given that $J_0(t) \Rightarrow p/\sqrt{p^2+1}$.

By (3) § 2·16

$$\int_t^\infty J_0(t)\, dt \Rightarrow 1 - 1/\sqrt{p^2+1} = (\sqrt{p^2+1}-1)/\sqrt{p^2+1}, \quad\ldots\ldots(1)$$

since by § 1·23 the integral with lower limit $t=0$ converges to the value unity.

---

* In orthodox form this would be written $cH(t)=c$. If the external p were dropped from (1), then in the symbolism of § 2·144, $cH(t) \Rightarrow c/p$, i.e. the L.T. of a constant would be a function of p.

★2·162.   Theorem 6a.—Function of $(a\sqrt{t^2-b^2})$, $t>b>0$.

If $f(a\sqrt{t^2-b^2}) \rightleftharpoons \phi(p)$ and $\displaystyle\int_b^\infty f(a\sqrt{t^2-b^2})\,dt$ is convergent,

then $\displaystyle\int_t^\infty f(a\sqrt{t^2-b^2})\,dt \rightleftharpoons e^{-pb} \lim_{p\to+0} [\phi(p)/p] - \phi(p)/p.$

Proof. By § 1·32, since the integral is convergent, we have

$$\int_b^\infty f(a\sqrt{t^2-b^2})\,dt = \lim_{p\to+0} [\phi(p)/p], \quad\dots\dots\dots(1)$$

so

$$\int_b^\infty f(a\sqrt{t^2-b^2})\,dt - \int_b^t f(a\sqrt{t^2-b^2})\,dt$$

$$\rightleftharpoons p\int_b^\infty e^{-pt}\left[\int_b^\infty f(a\sqrt{t^2-b^2})\,dt\right]dt - \phi(p)/p, \quad\dots\dots(2)$$

by § 2·151, the last member being the L.T. of the second.

Hence $\displaystyle\int_t^\infty f(a\sqrt{t^2-b^2})\,dt \rightleftharpoons e^{-pb} \lim_{p\to+0} [\phi(p)/p] - \phi(p)/p. \quad\dots\dots(3)$

★2·163. Example. Find the L.T. of $\displaystyle\int_t^\infty J_0(a\sqrt{t^2-b^2})\,dt$, given that

$$J_0(a\sqrt{t^2-b^2}) \rightleftharpoons pe^{-b\sqrt{p^2+a^2}}/\sqrt{p^2+a^2}, \quad t>b>0.$$

By (3) § 2·162

$$\int_t^\infty J_0(a\sqrt{t^2-b^2})\,dt \rightleftharpoons \frac{e^{-b(p+a)}}{a} - \frac{e^{-b\sqrt{p^2+a^2}}}{\sqrt{p^2+a^2}}, \quad\dots\dots\dots(1)$$

provided the integral converges when the lower limit is b. Using procedure similar to that in 2° § 8, Appendix III, the reader will readily confirm that this condition is satisfied.

2·17. Theorem 7. If $e^{-p_0t}f(t)$ is bounded in $t\geqslant 0$, then with p real $>p_0>0$,

$$t^n f(t) \rightleftharpoons (-1)^n p \frac{d^n}{dp^n}[\phi(p)/p].^*$$

Proof. By (1) § 1·11

$$\int_0^\infty e^{-pt}f(t)\,dt = \phi(p)/p. \quad\dots\dots\dots\dots\dots(1)$$

*★ See footnote to § 2·11.

Differentiating under the integral sign with respect to p, we obtain

$$\int_0^\infty e^{-pt}tf(t)\,dt = -\frac{d}{dp}\,[\phi(p)/p]. \qquad\ldots\ldots\ldots\ldots(2)$$

This procedure is valid since by § 1·21, the l.h.s. of (2) is *uniformly* convergent with respect to p in $p_1\leqslant p\leqslant p_2$, $p_1>p_0$. Hence

$$tf(t) \Rightarrow -p\frac{d}{dp}\,[\phi(p)/p]. \qquad\ldots\ldots\ldots\ldots\ldots(3)$$

By repeating the above procedure, we find ultimately that when $p>p_0$,

$$\int_0^\infty e^{-pt}t^n f(t)\,dt = (-1)^n\frac{d^n}{dp^n}\,[\phi(p)/p], \qquad\ldots\ldots\ldots\ldots(4)$$

so $$\qquad\qquad t^n f(t) \Rightarrow (-1)^n p\frac{d^n}{dp^n}\,[\phi(p)/p], \qquad\ldots\ldots\ldots\ldots(5)$$

since by § 1·21 the l.h.s. of (4) is u.c. for $n\geqslant1$.

*2·171. Modification of (5) § 2·17.   Using (1) § 2·144, if the condition in § 2·17 is satisfied, we find that

$$t^m f(t) \Rightarrow (-1)^m\frac{d^m}{dp^m}\,[\Phi(p)]. \quad (m\geqslant0). \qquad\ldots\ldots\ldots\ldots(1)$$

Applying this to (3) § 2·144, with $f^{(n)}(t)$ for $f(t)$ leads to

$$t^m\frac{d^n}{dt^n}f(t) = t^m f^{(n)}(t)$$

$$\Rightarrow (-1)^m\frac{d^m}{dp^m}\times\left[p^n\Phi(p)-\sum_{r=0}^{n-1}p^{n-r-1}f^{(r)}(0)\right], \quad (n>m) \;\ldots(2)$$

$$\Rightarrow (-1)^m\frac{d^m}{dp^m}\,[p^n\Phi(p)] \quad (m\geqslant n) \qquad\ldots\ldots\ldots\ldots\ldots\ldots(3)$$

provided the conditions in § 2·14 are satisfied. These formulae are needed for the solution of differential equations with variable coefficients in Chapter VI.

2·172. Example.   Find the L.T. of (a) $t\sinh t$, (b) $t\,e^{at}$, given that $\sinh t \Rightarrow p/(p^2-1)$, and $e^{at}\Rightarrow p/(p-a)$, a real.

(a) By (3) § 2·17

$$t\sinh t \Rightarrow -p\frac{d}{dp}\,[1/(p^2-1)]=2p^2/(p^2-1)^2. \qquad\ldots\ldots\ldots(1)$$

(b) $$\qquad t\,e^{at} \Rightarrow -p\frac{d}{dp}\,[1/(p-a)]=p/(p-a)^2. \qquad\ldots\ldots\ldots\ldots(2)$$

Alternatively by (2) § 2·13

$$e^{at} \cdot t \Rightarrow \left(\frac{p}{p-a}\right) \cdot \frac{1}{p-a} = p/(p-a)^2. \qquad \ldots\ldots\ldots(3)$$

Note that $e^{-p_0 t} \sinh t$ is bounded in $t \geqslant 0$ if $p_0 > 1$, whilst $t e^{-(p_0 - a)t}$ is bounded if $p_0 > a$.

*2·18. Theorem 8.—The addition theorem.

If $f_r(t) \Rightarrow \phi_r(p)$, then $\sum\limits_{r=0}^{m} f_r(t) \Rightarrow \sum\limits_{r=0}^{m} \phi_r(p)$. When m is finite the theorem is obvious, but when m is infinite we must have

$$p \int_0^\infty e^{-pt} \sum_{r=0}^\infty f_r(t)\,dt = p \sum_{r=0}^\infty \int_0^\infty e^{-pt} f_r(t)\,dt \Rightarrow \sum_{r=0}^\infty \phi_r(p). \quad \ldots(1)$$

If $\sum\limits_{r=0}^{\infty} f_r(t)$ is absolutely and uniformly convergent in $t \geqslant 0$,

(1) holds if p real > 0. For $p \int_0^\infty e^{-pt}\,dt$ is convergent and the general result in reference 3, § 176A, is applicable. In certain cases, e.g. in finding the L.T. of a function defined by a power series, this result is not sufficiently general. By applying § 176B, reference 3, we can show, however, that (1) is valid under the following conditions (see § 15, Appendix III) :

(a) $\sum\limits_{r=0}^{\infty} f_r(t)$ converges uniformly in the closed interval $0 \leqslant t \leqslant h$, $h > 0$, and

(b) either the *integral*

$$|p| \int_0^\infty |e^{-pt}| \sum_{r=0}^\infty |f_r(t)|\,dt, \qquad \ldots\ldots\ldots\ldots(2)$$

or the *series*

$$|p| \sum_{r=0}^\infty \int_0^\infty |e^{-pt}|\,|f_r(t)|\,dt, \qquad \ldots\ldots\ldots\ldots(3)$$

converges.* For instance if $f_r(t) = t^r/r!$, it is shown in 4° § 6, Appendix II, that the series is not u.c. *throughout* the range $t \geqslant 0$.

---

* In (2), (3) p is an unrestricted number. If $p = u + iv$,
$$e^{-pt} = e^{-ut}(\cos vt - i \sin vt),$$
so $|e^{-pt}| = e^{-ut}$. Conditions (a), (b), are *sufficient* for the validity of (1) § 2·18, but cases occur where the *necessary* conditions are less stringent. Certain non-u.c. series may be integrated term by term [see § 1·75, reference 16].

It is, however, u.꜃. in every *finite* interval $0 \leqslant t \leqslant h$. Hence if p is real and >0, (1) holds if the integral

$$p \int_0^\infty e^{-pt} \sum_{r=0}^\infty |f_r(t)| \, dt = p \int_0^\infty e^{-(p-1)t} \, dt \quad \ldots\ldots\ldots\ldots(4)$$

is convergent, i.e. if $p>1$. But if $f_r(t) = (-1)^r t^{2r}/r!$, (1) does not hold, as we shall see in § 2·183, since neither (2) nor (3) is satisfied.

*2·181. Example. Find the L.T. of $\sin t$ by expansion and term by term integration.

By (1) § 1·11

$$\sin t \rightleftharpoons p \int_0^\infty e^{-pt} \sin t \, dt = p \int_0^\infty e^{-pt} \sum_{r=0}^\infty (-1)^r \frac{t^{2r+1}}{(2r+1)!} \, dt, \quad \ldots(1)$$

which has the form considered in § 2·18, with

$$f_r(t) = (-1)^r t^{2r+1}/(2r+1)!$$

This series converges absolutely and *uniformly* in every finite interval $0 \leqslant t \leqslant h$ (see § 6, Appendix II). Thus

$$p \int_0^\infty e^{-pt} \sum_{r=0}^\infty |f_r(t)| \, dt = p \int_0^\infty e^{-pt} \sinh t \, dt$$

$$= \frac{p}{2} \int_0^\infty e^{-pt}(e^t - e^{-t}) \, dt, \quad \ldots\ldots\ldots(2)$$

and this integral converges if p is real and >1. Hence if $p>1$,

$$\sin t \rightleftharpoons p \sum_{r=0}^\infty (-1)^r \int_0^\infty e^{-pt} \frac{t^{2r+1}}{(2r+1)!} \, dt = \sum_{r=0}^\infty (-1)^r/p^{2r+1}, \quad \ldots(3)$$

so $\quad \sin t \rightleftharpoons p/(p^2+1)$. $\ldots\ldots\ldots\ldots\ldots\ldots\ldots\ldots\ldots\ldots\ldots\ldots\ldots\ldots\ldots\ldots(4)$

It may be remarked that if p real >0

$$p \int_0^\infty e^{-pt} \sin t \, dt = p/(p^2+1),$$

but for validity of the above analysis we must have $p>1$.

2·182. Dimensional equivalence of f(t) and φ(p). If in a physical application of integral (1) § 1·11, t has a certain dimension, that of p is to be regarded as the reciprocal of the dimension of t. When t represents time, p represents t^{-1}, so the index pt is dimensionless, as also is $p \, dt$. Consequently if $f(t)$ and

$\phi(p)$ are expansible, the conditions in § 2·18 being satisfied,* there is term by term correspondence, and the dimensions of such terms must be identical, e.g.

$$\sum_{r=0}^{\infty}(-1)^r\frac{t^{2r+1}}{(2r+1)!} \rightleftharpoons \sum_{r=0}^{\infty}\frac{(-1)^r}{p^{2r+1}}.\quad\ldots\ldots\ldots\ldots(1)$$

Dimensional equivalence is sometimes useful for rough-checking a L.T. Suppose the L.T. of a function has been found otherwise than by expansion, and a rough dimensional check is needed. Let $f(t)$ and $\phi(t)$ be expanded giving, say,

$$a_0+a_1t^2+a_2t^4+\ldots \rightleftharpoons \frac{b_0}{p}+\frac{b_1}{p^3}+\frac{b_2}{p^5}+\ldots.\quad\ldots\ldots\ldots\ldots(2)$$

Then if the constants a_r, b_r are dimensionless, the dimensions on the r.h.s. of (2) are erroneous, and it should read

$$b_0+\frac{b_1}{p^2}+\frac{b_2}{p^4}+\ldots.\quad\ldots\ldots\ldots\ldots\ldots(3)$$

★2·183. A divergent case. Find the L.T. of e^{-t^2}.

Expanding the exponential, we have

$$e^{-t^2}=1-t^2+\frac{t^4}{2!}-\frac{t^6}{3!}+\ldots,\quad\ldots\ldots\ldots\ldots(1)$$

which is an absolutely and uniformly convergent series in every closed interval of t. Taking the L.T. term by term, we get

$$1-\frac{2!}{1!p^2}+\frac{4!}{2!p^4}-\frac{6!}{3!p^6}+\ldots,\quad\ldots\ldots\ldots\ldots(2)$$

which diverges. Thus (3) § 2·18 is not satisfied. Also for (2) § 2·18 we have $p\int_0^{\infty}e^{-pt+t^2}\,dt$, $p>0$, which diverges, so (2) cannot be the L.T. of e^{-t^2}. It is the asymptotic expansion [12, p. 298] of

$$p\int_{p/2}^{\infty}e^{-x^2+p^2/4}\,dx=(p/2)\int_0^{\infty}e^{-v}\,dv/\sqrt{(p^2/4)+v}$$

$$=\int_0^{\infty}e^{-v}.dv/\sqrt{1+(4v/p^2)}.\quad\ldots\ldots\ldots(3)$$

---

* This usually means that : (a) $f(t)$ is expansible in a series of ascending (*positive*) powers of t which converges uniformly in every finite interval $0\leqslant t\leqslant h$; (b) $\phi(p)$ is expansible in an absolutely convergent series of inverse (*negative*) powers of p with p real >0, or $R(p)>0$. Since $f(at)\rightleftharpoons\phi(p/a)$, if a has the inverse dimension of t, both sides are dimensionless.

By (1) § 1·11 the L.T. we seek is

$$p \int_0^\infty e^{-pt-t^2} dt = p e^{p^2/4} \int_0^\infty e^{-(t+p/2)^2} dt, \quad \text{......}(4)$$

$$= p e^{p^2/4} \int_{p/2}^\infty e^{-y^2} dy, \quad \text{......}(5)$$

where $y = (t + p/2)$, so

$$e^{-t^2} = \frac{\sqrt{\pi}}{2} p e^{p^2/4} \operatorname{erfc}(p/2). \quad \text{......}(6)$$

From the identity of the first integral in (3) and the second in (5), we see that (2) is the *asymptotic* expansion of the r.h.s. of (6).

2·19. Theorem 9.—The impulse theorem. Let $f(t; \epsilon)$ be a function having the following properties: (a) it is positive, single-valued, and continuous when $0 \leqslant t \leqslant \epsilon$ *; (b) it is zero when $t > \epsilon$ and $t < 0$; (c) $\int_0^\epsilon f(t; \epsilon) dt = 1$. Then if $f(t; \epsilon) \rightleftharpoons \phi(p; \epsilon)$, $\phi(p; \epsilon) \to p$ as $\epsilon \to 0$.

Proof. By (2) § 1·11, we have

$$\phi(p; \epsilon) = p \int_0^\epsilon e^{-pt} f(t; \epsilon) dt, \quad \text{......}(1)$$

$$= p \int_0^\epsilon f(t; \epsilon) dt - p \int_0^\epsilon (1 - e^{-pt}) f(t; \epsilon) dt, \quad \text{......}(2)$$

$$= p - p \int_0^\epsilon (1 - e^{-pt}) f(t; \epsilon) dt. \quad \text{......}(3)$$

Now if p is real and > 0, as t increases from 0 to ϵ, $(1 - e^{-pt})$ increases from 0 to $(1 - e^{-p\epsilon})$. But $f(t; \epsilon) \geqslant 0$, so the integral in (3) gives,

$$0 \leqslant \int_0^\epsilon (1 - e^{-pt}) f(t; \epsilon) dt \leqslant (1 - e^{-p\epsilon}) \int_0^\epsilon f(t; \epsilon) dt = (1 - e^{-p\epsilon}). \quad \text{...}(4)$$

Since $(1 - e^{-p\epsilon}) \to +0$ as $\epsilon \to +0$, so also does the first integral in (4), and, therefore, that in (3). Hence as $\epsilon \to +0$

$$\phi(p; \epsilon) \to p, \quad \text{......}(5)$$

or $\quad f(t; \epsilon) \rightleftharpoons p$, approximately, $\quad \text{......}(6)$

* This simple property is used for clarity. The theorem can be proved if $f(t; \epsilon)$ changes sign and is finitely discontinuous in $0 \leqslant t \leqslant \epsilon$. When $\epsilon := 0, f(t; \epsilon)$ is infinitely discontinuous.

when ϵ is small. By virtue of property (c), with $\epsilon = 0$, $f(t)$ has infinite amplitude but infinitesimal duration. In practical problems ϵ cannot be zero, but it may be small enough for the use of (6) to give an adequate approximation.

The integral in (c) is defined to be the impulse due to the force $f(t \; ; \; \epsilon)$, and it is represented by the shaded area in Fig. 5a.

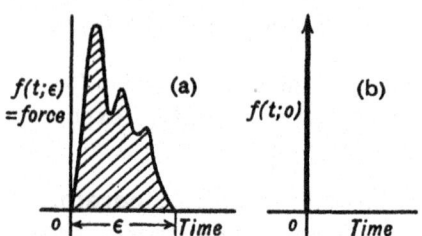

FIG. 5 (a).—Arbitrary form of impulsive force of duration $t = \epsilon$. The shaded area represents the *strength* of the ' impulse ', being unity for all values of ϵ.

(b).—Impulsive force of *infinite* amplitude but *unit strength*.

We shall use $I(t)$ to denote the limit as $\epsilon \to +0$ of $f(t \; ; \; \epsilon)$ defined above. Thus

$$\epsilon \to +0\, f(t \; ; \; \epsilon) = I(t) \Rightarrow p, \quad \dots\dots\dots\dots\dots(7)$$

signifies an impulsive force (mechanical, electrical, etc.) of *infinite* amplitude and *unit strength* which occurs at $t = 0$. If the impulse occurs at $t = h$, by § 2·12 we have

$$I(t - h) \Rightarrow e^{-ph}p. \quad \dots\dots\dots\dots\dots\dots(8)$$

When an impulsive force of this type is applied to a system, the energy is transferred thereto instantaneously. In a dissipative system, quiescent initially, the energy acquired commences to decay immediately after the occurrence of the impulse. Illustrations of the use of the impulsive function will be found in §§ 2·191, 2·251, 7·17 *et seq.*, 15–19, § 8·4, 9–13 § 8·5, and in Appendix VI.

Dimension of I (t). Since $I(t) \Rightarrow p$, it follows by § 2·182 that both $I(t)$ and p have dimension t^{-1}. In a physical problem it is essential to use the appropriate dimensions for an impulse, so we now take $\int_0^\epsilon f_1(t \; ; \; \epsilon)\,dt = A$. If f_1 has the dimensions of force, those of A are force × time. Moreover, we write

$$AI(t) = Ap, \quad\dots\dots\dots\dots\dots(9)$$

A having the dimensions of the applied force (electrical, mechanical, etc.) multiplied by time. $AI(t)$ represents an *impulsive force* of infinite amplitude, and A the *strength* of the impulse.

*2·191. Example. Evaluate $\displaystyle\int_0^\infty I(t-h)g(t)\,dt$, $h \geqslant 0$, if the function $g(t)$ is single-valued and continuous at $t = h$.

$I(t-h)$ is not a function in the ordinary sense. It is defined in a similar way to $I(t)$ in § 2·19 except that $(t-h)$ is written for t, the limits in the integral now being h and $(h+\epsilon)$. Thus

$$\int_h^{h+\epsilon} f(t-h\,;\ \epsilon)\,dt = 1 = \int_0^\epsilon f(\tau\,;\ \epsilon)\,d\tau. \quad\dots\dots\dots(1)$$

Then with τ for $(t-h)$

$$I = \int_0^\infty f(t-h\,;\ \epsilon)\,g(t)\,dt = \int_{-h}^\infty f(\tau\,;\ \epsilon)\,g(\tau+h)\,d\tau, \quad\dots\dots(2)$$

$$= \int_0^\epsilon f(\tau\quad,g(\tau+h)\,d\tau \quad\dots\dots\dots(3)$$

since $f(\tau\,;\ \epsilon)$ is zero when $\tau > \epsilon$ or < 0, $h \geqslant 0$, so the range of integration $\tau = (0,\ \epsilon)$ lies within $(-h,\ +\infty)$. But

$$\int_0^\epsilon f(\tau\,;\ \epsilon)\,g(\tau+h)\,d\tau$$

$$= g(h)\int_0^\epsilon f(\tau\,;\ \epsilon)\,d\tau + \int_0^\epsilon f(\tau\,;\ \epsilon)[g(\tau+h)-g(h)]\,d\tau, \quad\dots\dots(4)$$

$$= g(h) + \int_0^\epsilon f(\tau\,;\ \epsilon)[g(\tau+h)-g(h)]\,d\tau. \quad\dots\dots\dots(5)$$

Suppose η is the greatest value of $|\,g(\tau+h)-g(h)\,|$ in the range $0 \leqslant \tau \leqslant \epsilon$. Since $g(t)$ is continuous, η is finite and $\to 0$ with ϵ. Now $f(\tau\,;\ \epsilon) \geqslant 0$, so we have

$$0 \leqslant \int_0^\epsilon f(\tau\,;\ \epsilon)\,|\,g(\tau+h)-g(h)\,|\,d\tau \leqslant \eta\int_0^\epsilon f(\tau\,;\ \epsilon)\,d\tau = \eta, \quad\dots(6)$$

and, therefore, the integral on the r.h.s. of (5)$\to 0$ with ϵ.   Consequently

$$\int_0^\infty I(t-h)g(t)\,dt = g(h), \quad\ldots\ldots\ldots\ldots(7)^*$$

this being the value of $g(t)$ at the point where the impulse occurs.   Since $I(t-h)$ is zero except at $t=h$, the limits of integration may be written $(h-\epsilon_1)$ and $(h+\epsilon_2)$, $\epsilon_1 \to 0$, $\epsilon_2 \to 0$. Formula (7) is applied in 15, 17, 18, 19 § 8·4, and in Appendix VI.

*2·192. Inversion of p^ν when $R(\nu) > 1$.

In § 2·19 it is shown that under certain conditions p may be inverted as $I(t)$, the impulsive function of infinite amplitude but unit strength.   So far as the Laplace transform is concerned, a *satisfactory* inversion of p^ν, $R(\nu) > 1$ does not seem to have been found.   This being so, we ask whether a transform expansible in ascending powers of p can be inverted term by term?   The answer is No!   An example will serve to demonstrate the point in question.   From the list we obtain

$$e^{-1/t}/\sqrt{t} \rightleftharpoons \sqrt{\pi p}\, e^{-2\sqrt{p}}. \quad\ldots\ldots\ldots\ldots\ldots(1)$$

The expansion of both members of (1) is absolutely convergent, but except for the first, there is no term by term correspondence. †   Thus

$$e^{-1/t}t^{-1/2} = t^{-1/2} - t^{-3/2} + \frac{t^{-5/2}}{2!} - \frac{t^{-7/2}}{3!} + \ldots , \quad\ldots\ldots(2)$$

and $$(\pi p)^{1/2}e^{-2\sqrt{p}} = \pi^{1/2}\left[p^{1/2} - 2p + \frac{4p^{3/2}}{2!} - \frac{8p^2}{3!} + \ldots\right]. \quad\ldots\ldots(3)$$

The L.T. of $t^{-1/2}$ is $(\pi p)^{1/2}$, i.e. the first term on the r.h.s. of (3), but by § 1·12, the remainder of the terms in (2) do not have

---

* This result may be obtained also—but less rigorously—by applying the product theorem § 2·24.   Since the impulse exists at $t=h$ only, we may write $I(t-h) = I(h-t)$.   With $g(t) = \phi_1(p)$, $I(t) = \phi_2(p) = p$, we get $\phi_1\phi_2/p = \phi_1(p) = g(t)$.   Hence by (5) § 2·24

$$\int_0^\infty I(h-\lambda)g(\lambda)d\lambda = \int_0^h I(h-\lambda)g(\lambda)d\lambda = g(h),$$

where h is written for t.   Another procedure is given in § 2·25.

† The expansions in (2), (3) do not comply with the footnote to § 2·182, since (2) is in negative and (3) in positive powers, which is the wrong way round.

Laplace transforms, since their indices are all < -1. The second term in (3) may be inverted as $-2\pi^{1/2}I(t)$, but those which follow cannot be inverted, because they are not L.T.S. as defined by (1) § 1·11.

The difficulty experienced here can be surmounted by recourse to complex integration using the contour on the left of Fig. 18. This, however, is beyond the scope of the text, but can be treated by aid of reference 12.

2·193. Remarks on inverting the order of integration. In most theorems which follow, the proofs depend mainly upon inversion of the order of integration of infinite repeated integrals.

Sufficient conditions to be satisfied are set out in C_1, C_2, § 9 Appendix III. In some cases the conditions may be too narrow, broader ones being permissible.

2·21. Theorem 10. $\dfrac{f(t)}{t} = p \displaystyle\int_p^\infty \dfrac{\phi(p)}{p}\,dp. \quad (p > 0).$

Proof. By definition (1) § 1·11

$$\int_0^\infty e^{-pt}f(t)\,dt = \phi(p)/p, \quad\quad\quad\quad\dots\dots\dots(1)$$

so $\quad p\displaystyle\int_p^\infty \left[\int_0^\infty e^{-pt}f(t)\,dt\right]dp = p\int_p^\infty \dfrac{\phi(p)}{p}\,dp, \quad\dots\dots(2)$

If the order of integration on the l.h.s. of (2) may be inverted,

$$p\int_0^\infty e^{-pt}\frac{f(t)}{t}\,dt = p\int_p^\infty \frac{\phi(p)}{p}\,dp, \quad\quad\dots\dots(3)$$

so $\quad\quad\quad\quad \dfrac{f(t)}{t} = p\displaystyle\int_p^\infty \dfrac{\phi(p)}{p}\,dp. \quad\quad\dots\dots(4)$

The conditions for the changed order of integration in (2) are given in § $9C_1$, Appendix III.

Corollary. Applying (3) § 2·15 to (4), we get

$$\int_0^t \frac{f(t)}{t}\,dt = \int_p^\infty \frac{\phi(p)}{p}\,dp, \quad\quad\quad\dots\dots\dots(5)$$

provided that $\lim\limits_{t\to+\infty} e^{-pt}\displaystyle\int_0^t f(\tau)\,d\tau \to 0$, and § $9C_1$, Appendix III, is satisfied.

2·211. Example.   Find the L.T. of $\sin t/t$.

From the list and (4) § 2·21

$$\frac{\sin t}{t} = p\int_p^\infty \frac{dp}{p^2+1} = p\,[\tfrac{1}{2}\pi - \tan^{-1}p], \quad\ldots\ldots\ldots(1)$$

$$= p\,\tan^{-1}(1/p). \quad\ldots\ldots\ldots(2)$$

By § $9C_1$, Appendix III, or (3), (6) $2°$ § 10, Appendix III, the order of integration in (2) § 2·21 is invertible.   Theorem 10 is inapplicable to functions like $\sin t/t$, $\cos t$, $J_0(t)$, owing to the infinite discontinuity at the origin in each case.   It is applicable to $J_\nu(t)$, $R(\nu) > 0$, since C_1 is satisfied.

2·212. Example.   Find the L.T. of $\int_0^t \frac{\sin t}{t}\,dt$.

By (4), (5) § 2·21 and (2) § 2·211, or by (3) § 2·15 and (2) § 2·211

$$\int_0^t \frac{\sin t}{t}\,dt = \tan^{-1}(1/p), \quad\ldots\ldots\ldots(1)$$

since $\lim\limits_{t\to+\infty} e^{-pt}\int_0^t \sin\tau\,d\tau \to 0$, $\int_0^t \sin\tau\,d\tau$ being oscillatory but bounded as $t \to +\infty$.

2·22. Theorem II.—The infinite integral theorem.

$$\int_0^\infty \frac{f(t)}{t}\,dt = \int_0^\infty \frac{\phi(p)}{p}\,dp = \text{a constant}.$$

Proof.   Taking the lower limit of the p integrals in (2) § 2·21 as zero, we get

$$\int_0^\infty \left[\int_0^\infty e^{-pt} f(t)\,dt\right] dp = \int_0^\infty \frac{\phi(p)}{p}\,dp. \quad\ldots\ldots\ldots(1)$$

If the order of integration may be inverted, the above result follows on evaluating the p integral on the l.h.s.   The conditions for the changed order of integration in (1) are given in § $9C_1$, Appendix III.

2·221. Example.   Evaluate $\int_0^\infty \frac{\sin t}{t}\,dt$.

By § 2·22 and the list,

$$\int_0^\infty \frac{\sin t}{t}\,dt = \int_0^\infty \frac{dp}{p^2+1}, \quad\ldots\ldots\ldots(1)$$

$$= \left[\tan^{-1}p\right]_0^\infty = \tfrac{1}{2}\pi, \quad\ldots\ldots\ldots(2)$$

provided the order of integration in (1) § 2·22 is invertible. This is shown to be permissible in 2° § 10, Appendix III.

Unless complex integration is used, direct evaluation of the integral on the l.h.s. of (1) is a protracted process. Although the foregoing procedure is delightfully terse and simple, it does not represent the complete picture, since proof of the inversion must be included. Cases like this, *where the answer is known already*, must be viewed in proper perspective!

2·222. Example. Evaluate $\int_0^\infty \dfrac{J_3(t)}{t}\,dt$.

By § 2·22 and the list

$$\int_0^\infty \frac{J_3(t)}{t}\,dt = \int_0^\infty \frac{dp}{\sqrt{p^2+1}\,(p+\sqrt{p^2+1})^3}, \quad\quad\quad\quad\quad (1)$$

$$= \int_0^\infty \frac{d\,(p+\sqrt{p^2+1})}{(p+\sqrt{p^2+1})^4} = -\left[\frac{1}{3\,(p+\sqrt{p^2+1})^3}\right]_0^\infty, \quad \ldots (2)$$

$$= 1/3, \quad\quad\quad\quad\quad\quad\quad\quad\quad\quad\quad\quad\quad\quad (3)$$

since § $9C_1$, Appendix III, is satisfied.

2·23. Theorem 12. $\displaystyle\int_t^\infty \frac{f(t)}{t}\,dt = \int_0^p \frac{\phi(p)}{p}\,dp, \quad t > 0.$

Proof. Multiplying both sides of (3) § 2·11 by da/a and integrating from $a = 1$ to $+\infty$, we get

$$p \int_1^\infty \left[\int_0^\infty e^{-pt} f(at)\,dt\right] \frac{da}{a} = \int_1^\infty \phi(p/a)\frac{da}{a}. \quad\quad\ldots (1)$$

Writing $x = at$, $p/a = y$, and assuming that the order of integration on the l.h.s. is invertible, we obtain

$$p \int_0^\infty e^{-pt}\left[\int_t^\infty \frac{f(x)}{x}\,dx\right] dt = \int_0^p \frac{\phi(y)}{y}\,dy, \quad\quad\ldots (2)$$

or

$$\int_t^\infty \frac{f(t)}{t}\,dt = \int_0^p \frac{\phi(p)}{p}\,dp. \quad\quad\ldots (3)*$$

The conditions for the changed order of integration in (1) are given in § $9C_2$, Appendix III.

* If (5) § 2·21 and theorem 11 are applicable (3) follows immediately.

2·231. Example. Find the L.T.S. of

$$(a) \ \mathrm{si}\,(t) = -\int_t^\infty \frac{\sin x}{x}\,dx,$$

$$(b) \ \mathrm{ci}\,(t) = -\int_t^\infty \frac{\cos x}{x}\,dx, \quad t>0.$$

(a) By (3) § 2·23 and the list, we have

$$\mathrm{si}\,(t) \bumpeq -\int_0^p \frac{dp}{p^2+1} = -\tan^{-1} p, \quad \dots\dots\dots\dots(1)$$

$$= \tan^{-1}(1/p) - \tfrac{1}{2}\pi. \quad \dots\dots\dots\dots(2)$$

(b) In like manner

$$\mathrm{ci}\,(t) \bumpeq -\int_0^p \frac{p\,dp}{p^2+1} = -\tfrac{1}{2}\Big[\log(p^2+1)\Big]_0^p, \quad \dots\dots(3)$$

$$= \log\,(1/\sqrt{p^2+1}). \quad \dots\dots\dots\dots\dots(4)$$

In both (a), (b), inversion of the order of integration in (1) § 2·23 is permissible, since conditions § 9 C_2, Appendix III are satisfied.

2·24. Theorem 13.—The product theorem. If $e^{-p_0 t}f_1(t)$, $e^{-p_0 t}f_2(t)$ are bounded in $t\geqslant 0$, then

$$\phi_1\phi_2/p \bumpeq \int_0^t f_1(t-\lambda)f_2(\lambda)d\lambda = \int_0^t f_1(\lambda)f_2(t-\lambda)\,d\lambda,$$

where $f_1(t) \bumpeq \phi_1(p)$ and $f_2(t) \bumpeq \phi_2(p)$.

Proof. By § 1·21 the integrals

$$p\int_0^\infty e^{-pu}f_1(u)\,du = \phi_1(p), \ \text{and} \ p\int_0^\infty e^{-pv}f_1(v)\,dv = \phi_2(p), \ \dots(1)$$

are *absolutely* and uniformly convergent in $p>p_0>0$. Absolute convergence implies that the integral of the product is equal to the product of the integrals. Thus

$$p\int_0^\infty\int_0^\infty e^{-p(u+v)}f_1(u)f_2(v)\,du\,dv = \phi_1\phi_2/p, \quad \dots\dots\dots(2)$$

the integration being taken over the quadrant where $u>0$, $v>0$, as shown in Fig. 6a. Substituting $u=t-\lambda$, $v=\lambda$ in (2) leads to

$$p\int_0^\infty\int_0^t e^{-pt}f_1(t-\lambda)f_2(\lambda)\,d\lambda\,dt = \phi_1\phi_2/p, \quad \dots\dots\dots(3)$$

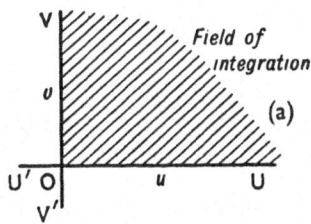
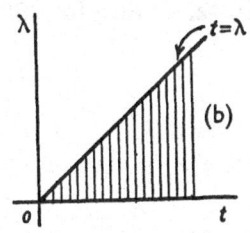

FIG. 6 (a), (b).—Diagrams illustrating field of integration used in deriving the 'product theorem'.

the integration being over the region between the t axis and the straight line $t = \lambda$ in the (t, λ) plane as shown in Fig. 6b. Now (3) may be written in the form

$$p \int_0^\infty e^{-pt} \left[\int_0^t f_1(t - \lambda) f_2(\lambda) \, d\lambda \right] dt = \phi_1 \phi_2/p, \quad \ldots\ldots(4)$$

from which it follows that

$$\int_0^t f_1(t - \lambda) f_2(\lambda) \, d\lambda \rightleftharpoons \phi_1 \phi_2/p. \quad \ldots\ldots\ldots(5)$$

By symmetry it is obvious that

$$\int_0^t f_1(\lambda) f_2(t - \lambda) \, d\lambda \rightleftharpoons \phi_1 \phi_2/p. \quad \ldots\ldots\ldots(6)$$

The criterion is *absolute* convergence of the integrals in (1). This is ensured by boundedness of $e^{-p_0 t} f_1(t)$, $e^{-p_0 t} f_2(t)$.

The theorem holds also under the extended conditions given in (1) § 1·211, i.e., if the integrals in (1) above are absolutely convergent.

2·241. Theorem 13a.

When the variable in § 2·24 takes the form $a\sqrt{t^2 - b^2}$ in one of the functions, we have

$$\int_b^t f_1(t - \lambda) f_2(a\sqrt{\lambda^2 - b^2}) \, d\lambda \rightleftharpoons \phi_1 \phi_2/p, \quad \ldots\ldots\ldots(1)$$

with $f_1(t) \rightleftharpoons \phi_1(p)$, and $f_2(a\sqrt{t^2 - b^2}) \rightleftharpoons \phi_2(p)$. Formula (1) is used in § 4·332.

2·242. Example. Show that

$$\int_0^t J_0(\lambda) J_1(t-\lambda) d\lambda = J_0(t) - \cos t.$$

Both $J_0(t)$ and $J_1(t)$ are bounded in $t \geqslant 0$, so the conditions for the validity of § 2·24 are satisfied. From the list,

$$J_0(t) \rightleftharpoons p/\sqrt{p^2+1} = \phi_1, \quad \dots \dots \dots \dots \dots \dots \dots \dots \dots \dots (1)$$

and $J_1(t) \rightleftharpoons p/\sqrt{p^2+1}\,(p + \sqrt{p^2+1}) = p(\sqrt{p^2+1} - p)/\sqrt{p^2+1}$,

when the numerator and denominator are each multiplied by $(\sqrt{p^2+1} - p)$.

Thus $\qquad J_1(t) \rightleftharpoons (p - p^2/\sqrt{p^2+1}) = \phi_2. \quad \dots \dots \dots \dots (2)$

Hence by (5) § 2·24, (1) and (2) above,

$$\int_0^t J_0(\lambda) J_1(t-\lambda) d\lambda \rightleftharpoons \frac{p}{\sqrt{p^2+1}} - \frac{p^2}{p^2+1} \rightleftharpoons J_0(t) - \cos t, \quad \dots(3)$$

as a consequence of Lerch's theorem § 1·16, both functions on the r.h.s. being continuous in $t \geqslant +0$.

2·243. Differentiation of (5), (6) § 2·24. If $f_1(t)$ and $f_2(t)$ are continuous in the range of integration, then by aid of (1), § 12A, Appendix III, we get

$$\frac{d}{dt}\int_0^t f_1(\lambda)f_2(t-\lambda)d\lambda = f_1(t)f_2(0) + \int_0^t f_1(\lambda)f_2'(t-\lambda)d\lambda \rightleftharpoons \phi_1\phi_2, \quad \dots(1)$$

$$= f_1(t)f_2(0) + \int_0^t f_1(t-\lambda)f_2'(\lambda)d\lambda \rightleftharpoons \phi_1\phi_2, \quad \dots(2)$$

the latter integral being obtained by writing $t - \lambda = \mu$ in the r.h.s. of (1) and then changing μ for λ. Also

$$\frac{d}{dt}\int_0^t f_1(t-\lambda)f_2(\lambda)d\lambda = f_1(0)f_2(t) + \int_0^t f_1'(t-\lambda)f_2(\lambda)d\lambda \rightleftharpoons \phi_1\phi_2, \quad \dots(3)$$

$$= f_1(0)f_2(t) + \int_0^t f_1'(\lambda)f_2(t-\lambda)d\lambda \rightleftharpoons \phi_1\phi_2. \quad \dots(4)$$

The identity of the results (1)–(4) should be noted, i.e. the derivative of the l.h.s. of (5) § 2·24 or of (6) § 2·24 may be expressed in four different ways.

*2·244. Example. Solve the Poisson integral equation

$$f(t) = f_1(t) \pm \int_0^t f(\lambda) f_2(t - \lambda) d\lambda$$

for $f(t)$, where $f_1(t)$, $f_2(t)$ are known functions having L.T.S. Take $f(t) \Rightarrow \phi(p)$, i.e. we assume $f(t)$ to have a L.T., so that $f_1(t) \Rightarrow \phi_1(p)$, and $f_2(t) \Rightarrow \phi_2(p)$. Then by § 2·24 we have

$$\phi = \phi_1 \pm \phi\phi_2/p, \quad \dots\dots\dots(1)$$

so
$$\phi = p\phi_1/(p \mp \phi_2). \quad \dots\dots\dots(2)$$

Thus
$$p\int_0^\infty e^{-pt}f(t)dt = p\phi_1/(p \mp \phi_2), \quad \dots\dots\dots(3)$$

and $f(t)$ is found if this integral equation can be solved. In certain cases the solution can be obtained by aid of the list. If this fails, the simplest procedure is to apply the Mellin theorem in Appendix IV, provided that the requisite conditions are satisfied. Then

$$f(t) = \frac{1}{2\pi i}\int_{c-i\infty}^{c+i\infty} e^{zt}\phi_1(z)dz/[z \mp \phi_2(z)]. \quad \dots\dots\dots(4)$$

$f(t)$ and $f_2(t)$ must fulfil the conditions for validity of § 2·24, either as stated there or in its extended form (see end of § 2·24).

It may be remarked that if A, B, C are constants, the equation

$$Af(t) = Bf_1(t) \pm C\int_0^t f(\lambda)f_2(t - \lambda)\,d\lambda, \quad \dots\dots\dots(5)$$

may be solved for $f(t)$ by the above procedure.

*2·245. Illustration of equation in § 2·244. Let

$$f_1(t) = \cos t, \quad f_2(t) = J_1(t),$$

then
$$\phi_1 = p^2/(p^2 + 1), \quad \dots\dots\dots(1)$$

and
$$\phi_2 = p - p^2/\sqrt{p^2 + 1}, \quad \dots\dots\dots(2)$$

by (2) § 2·242. Substituting in (2) § 2·244, with the upper sign we obtain

$$\phi = \left(\frac{p^3}{p^3 + 1}\right) \cdot \frac{\sqrt{p^2 + 1}}{p^2}, \quad \dots\dots\dots(3)$$

$$= p/\sqrt{p^2 + 1}, \quad \dots\dots\dots(4)$$

so by the list $\qquad f(t) = J_0(t).$ ....(5)

Consequently $\qquad J_0(t) = \cos t + \int_0^t J_0(\lambda) J_1(t - \lambda) d\lambda,$ ....(6)

which is merely (3) § 2·242.

*2·246. Example. Solve Abel's integral equation :

$$f(t) = \int_0^t f_1(\lambda) d\lambda / (t - \lambda)^\nu, \qquad\qquad ....(1)$$

where $f(t)$ is known but $f_1(t)$ unknown ; $f(0) = 0,$ $0 < \nu < 1.$

Take $f(t) \Rightarrow \phi, f_1(t) \Rightarrow \phi_1, f_2(t) = t^{-\nu} \Rightarrow \Gamma(1 - \nu) p^\nu = \phi_2.$

Then by (6), § 2·24,

$$\phi = \phi_1 \phi_2 / p = \phi_1 \Gamma(1 - \nu) p^{\nu-1}, \qquad\qquad ....(2)$$

so $\qquad \phi_1 = \phi / \Gamma(1 - \nu) p^{\nu-1} = \dfrac{1}{\Gamma(\nu)\Gamma(1 - \nu) p} \left[p\phi \left\{ \dfrac{\Gamma(\nu)}{p^{\nu-1}} \right\} \right].$ ......(3)

Since $f(0) = 0,$ $f'(t) \Rightarrow p\phi$ by (10) § 2·14. Thus using the relationship $\Gamma(\nu)\Gamma(1 - \nu) = \pi / \sin \nu\pi,$ and applying (6) § 2·24 to (3), we obtain the well-known solution :

$$f_1(t) = \frac{\sin \nu\pi}{\pi} \int_0^t f'(\lambda) (t - \lambda)^{\nu-1} d\lambda. ....(4)$$

(4) is the inverse of (1).

Abel's equation dates back to 1823 (see his *Œuvres*), and its solution was the first instance of the *inversion* of a definite integral—an epoch-making step. The problem in which it arose may be stated as follows : A mass m, starting from rest at O, moves down a frictionless chute mounted in a vertical plane. If the time taken to fall a vertical distance h below O is $f(h)$, a *given* function of h, find the curve of the chute. In solving this problem, Abel obtained (1) in the form

$$f(h) = \frac{1}{\sqrt{2g}} \int_0^h \frac{f_1(z) dz}{\sqrt{h - z}} \cdot \qquad\qquad ....(5)$$

Here $\nu = 1/2.$ If $f(h)$ is a constant, the curve is, of course, a cycloid.

It may be remarked that (1) is a particular case of (5), § 2·244, with $A = 0,$ $B = -1,$ $\pm C = 1.$

*2·25. Theorem 14. If
$$f(t) \rightleftharpoons \phi(p) \text{ and } \xi(x, t) \rightleftharpoons \phi_1(p)h(p)e^{-x\,h(p)},$$

then $\displaystyle\int_0^\infty \xi(x, t)f(x)dx \rightleftharpoons \phi_1(p)\phi[h(p)]$, where $\phi_1(p)$, $h(p)$ are continuous functions of p, independent of x, $R[h(p)] \geqslant p_0 > 0$, x real and $\geqslant 0$.

Proof. Writing $h(p)$ for p in (1) § 1·11, we get

$$\phi[h(p)] = h(p)\int_0^\infty e^{-t\,h(p)}f(t)dt. \quad \dots\dots\dots\dots(1)$$

Since $f(t) \rightleftharpoons \phi(p)$, $\displaystyle\int_0^\infty e^{-pt}f(t)dt$ converges for $R(p) \geqslant p_0$, so (1) converges if $R[h(p)] \geqslant p_0 > 0$. Multiplying both sides of (1) by $\phi_1(p)$ and replacing t by x leads to

$$\phi_1(p)\phi[h(p)] = \int_0^\infty \phi_1(p)h(p)e^{-x\,h(p)}f(x)dx. \quad \dots\dots(2)$$

By hypothesis

$$p\int_0^\infty e^{-pt}\xi(x, t)dt = \phi_1(p)h(p)e^{-x\,h(p)}. \quad \dots\dots\dots(3)$$

Substituting from (3) into (2) gives

$$\phi_1(p)\phi[h(p)] = p\int_0^\infty e^{-pt}\left[\int_0^\infty \xi(x, t)f(x)dx\right]dt, \quad \dots\dots(4)$$

provided that the order of integration may be inverted. Then

$$\int_0^\infty \xi(x, t)f(x)dx \rightleftharpoons \phi_1(p)\phi[h(p)]. \quad \dots\dots\dots(5)$$

Sufficient conditions for the changed order of integration are either

C_1.   (a) $f(x)$ is continuous in $x \geqslant 0$;

   (b) $\xi(x, t)$ is continuous in $x \geqslant 0$, and in $t \geqslant 0$;

   (c) $\displaystyle\int_0^\infty e^{-pt}\xi(x, t)f(x)dx$ is u.c. in $t \geqslant 0$, so $\displaystyle\int_0^\infty \xi(x, 0)f(x)dx$ is convergent ;

   (d) either

$$\int_0^\infty e^{-pt}dt\int_0^\infty \xi(x, t)f(x)dx \text{ or } \int_0^\infty f(x)dx\int_0^\infty e^{-pt}\xi(x, t)dt$$

is absolutely convergent ;

or C_2.   (a), (b) as in C_1 ;

(c) As in (c) at C_1, but in the range $0 \leqslant t \leqslant h$;

(d) $\displaystyle\int_0^\infty f(x)dx \int_0^t e^{-p\tau}\xi(x,\tau)d\tau$ is u.c. in $t \geqslant 0$.

When $\xi(x,t)$ takes the form $\xi(t-x)$ or $\xi(x,t-x)$, the r.h.s. of (3) contains the factor e^{-px}, e.g. if $h(p)=p$, and the lower limit in (4) is $x \geqslant 0$. Since $x < t$, the upper limit in (5) is now t. Conditions for the inverted order of integration are, by § 9A, Appendix III :

C_3.   (a) $f(x)$ is continuous in $0 \leqslant x \leqslant t$;

(b) $\xi(t-x)$ or $\xi(x,t-x)$ is continuous in $0 \leqslant x \leqslant t$;

(c) $\displaystyle\int_x^\infty e^{-pt}\begin{Bmatrix}\xi(t-x)\\\xi(x,t-x)\end{Bmatrix}f(x)dt$ is u.c. with respect to x in $0 \leqslant x \leqslant t$.

*2·251.   Particular cases of § 2·25.

1°. When $\phi_1(p)=1$, (5) § 2·25 degenerates to

$$\int_0^\infty \xi(x,t)f(x)dx \rightleftharpoons \phi[h(p)]. \qquad \dots\dots\dots\dots(1)$$

2°. When $h(p)=p$, (1) takes the form

$$\int_0^t \xi(t-x)f(x)dx \rightleftharpoons \phi(p), \qquad \dots\dots\dots\dots(2)$$

so

$$\int_0^t \xi(t-x)f(x)dx = f(t), \qquad \dots\dots\dots\dots(3)$$

if $f(x)$ is single-valued and continuous at $x=t$. Thus by (7) § 2·191

$$\xi(t-x)=I(t-x), \qquad \dots\dots\dots\dots(4)$$

the impulsive function $I(t)$ at $x=t$. This result can be obtained from (3) § 2·25 also, because

$$p\int_0^\infty e^{-pt}\xi(t-x)dt = pe^{-px}; \qquad \dots\dots\dots\dots(5)$$

so by (8) § 2·19, we get (4).

3°. Writing $\phi_1(p)/p$ for $\phi_1(p)$ and $h(p)=p$ in (3) § 2·25 leads to $\xi(t-x) \rightleftharpoons e^{-xp}\phi_1(p)$, so $\xi(t) \rightleftharpoons \phi_1(p)$. Substituting in (5) § 2·25 gives

$$\int_0^t \xi(t-x)f(x)dx \rightleftharpoons \phi_1\phi/p, \qquad \dots\dots\dots\dots(6)$$

the upper limit being t, since $x < t$. Thus the product theorem § 2·24 is a particular case of § 2·25. The latter is used to derive formulae of the type on p. 12, reference 13. After perusing §§ 2·252, 2·253, these formulae should be obtained by the reader as an exercise.

★2·252. Example. Show that

$$\int_0^t J_0[2\sqrt{x(t-x)}]f(x)dx = [p/(p^2+1)]\phi[(p^2+1)/p].$$

In (5) § 2·25 take $h(p) = (p^2+1)/p$, $\phi_1(p) = p/(p^2+1)$, and (3) § 2·25 gives

$$p\int_x^\infty e^{-pt}\xi(x, t)dt = e^{-x(p+1/p)}, \quad\quad\quad\text{......(1)}$$

so $$\xi(x, t) = J_0[2\sqrt{x(t-x)}], \quad\quad\text{......(2)}$$

$t > x$,★ by the list and (3) § 2·12. Hence by (5) § 2·25 we obtain

$$\int_0^t J_0[2\sqrt{x(t-x)}]f(x)dx = [p/(p^2+1)]\phi[(p^2+1)/p]. \quad\text{......(3)}$$

In (2) $\xi(x, t)$ has the form $\xi(x, t-x)$, and condition (b) in C_3 is satisfied. Hence for (3) to be true, $f(x)$ must be such that conditions (a), (c) are satisfied also.

If (p^2-1) is written for (p^2+1) in (3) $R(p) > 1$, and (a), (c) in C_3 are satisfied, then

$$\int_0^t I_0[2\sqrt{x(t-x)}]f(x)dx = [p/(p^2-1)]\phi[(p^2-1)/p]. \quad\text{...(4)}$$

★2·253. Example. Show that

$$1/(2^{\frac{1}{2}n}t^{\frac{1}{2}(n+1)}\sqrt{\pi})\int_0^\infty e^{-x^2/4t}He_n(x/\sqrt{2t})f(x)dx = p^{n/2}\phi(p^{1/2}).$$

In (3) § 2·25 let $\phi_1(p) = p^{n/2}$ and $h(p) = p^{1/2}$. Then

$$\xi(x, t) = p^{\frac{1}{2}(n+1)}e^{-xp^{1/2}}, \quad\quad\quad\text{......(1)}$$

and from the list we obtain

$$\xi(x, t) = e^{-x^2/4t}He_n(x/\sqrt{2t})/(2^{\frac{1}{2}n}t^{\frac{1}{2}(n+1)}\sqrt{\pi}). \quad\quad\text{......(2)}$$

Whence by (5) § 2·25 the above result follows immediately, provided that one of the sets of conditions C_1, C_2, is satisfied. This depends upon $f(x)$. $He_n(u)$ is Hermite's polynomial, of order n [13].

★ Since (1) has the factor e^{-px}.

If $n=0$, we have the particular case

$$(1/\sqrt{\pi t})\int_0^\infty e^{-x^2/4t}f(x)\,dx \rightleftharpoons \phi(p^{1/2}). \dots\dots\dots(3)$$

*2·254. Example. Show that

$$\int_0^t I_0(a\sqrt{t^2-x^2})f(x)\,dx \rightleftharpoons \frac{p}{(p^2-a^2)}\phi(\sqrt{p^2-a^2}).$$

In (3) § 2·25 let $\phi_1(p)=p/(p^2-a^2)$, and $h(p)=\sqrt{p^2-a^2}$.

Then
$$\xi(x, t) \rightleftharpoons \frac{p}{\sqrt{p^2-a^2}}\cdot e^{-x\sqrt{p^2-a^2}}, \quad \dots\dots\dots\dots(1)$$

and from the list we find that

$$\xi(x, t) = I_0(a\sqrt{t^2-x^2}). \quad\dots\dots\dots\dots\dots(2)$$

Hence by (5) § 2·25, if the appropriate conditions for inverting the order of integration are satisfied, it follows that

$$\int_0^t I_0(a\sqrt{t^2-x^2})f(x)\,dx \rightleftharpoons \frac{p}{(p^2-a^2)}\phi(\sqrt{p^2-a^2}),\dots\dots(3)$$

the upper limit being t, since $t>x$. The application of this theorem to the cable problem in § 4·321 shortens the analytical work and enables the result to be expressed concisely.

If for a we write ia, we obtain the theorem

$$\int_0^t J_0(a\sqrt{t^2-x^2})f(x)\,dx \rightleftharpoons \frac{p}{p^2+a^2}\phi(\sqrt{p^2+a^2}), \quad\dots\dots(4)$$

provided the requisite conditions are satisfied.

*2·31. Approximations.

1°. Suppose $f(t)$ can be represented by a convergent series such that

$$f(t)=\sum_{r=0}^\infty a_r t^{r+\nu} \rightleftharpoons \sum_{r=0}^\infty a_r \Gamma(\nu+r+1)p^{-r-\nu}, \quad R(\nu)>-1,$$

there being term by term correspondence. When t is small enough, but positive, we may write

$$f(t)\simeq a_0 t^\nu \rightleftharpoons a_0\Gamma(\nu+1)p^{-\nu},$$

provided the remaining terms on the r.h.s. above may be neglected. This condition is satisfied if p is made large enough. Then the major part of the integral (1) § 1·11 is obtained in the neighbourhood of the origin. Thus by making p large, and inverting the L.T., the form of $f(t)$ can be ascertained when t is small.

Example. What is the form of (a) the Bessel function $J_0(t)$, (b) Struve's function $H_0(t)$ [11] in the vicinity of the origin.

(a) $$J_0(t) \rightleftharpoons p/\sqrt{p^2 + 1} \rightarrow 1 \text{ as } p \rightarrow +\infty, \quad \dots\dots\dots(1)$$

so $$J_0(t) \rightarrow 1 \text{ as } t \rightarrow +0. \quad \dots\dots\dots\dots\dots(2)$$

Here we may let $p \rightarrow +\infty$, since $J_0(t) \rightarrow$ a constant. For a closer approximation we take

$$p/\sqrt{p^2 + 1} = 1/[1 + (1/p^2)]^{1/2} \simeq 1 - (1/2p^2), \quad \dots\dots(3)$$

$$\simeq 1 - t^2/2^2, \quad \dots\dots\dots\dots\dots(4)$$

if p is large enough.

(b) $$H_0(t) \rightleftharpoons \frac{2p}{\pi\sqrt{p^2 + 1}} \log\left(\frac{1 + \sqrt{p^2 + 1}}{p}\right), \quad \dots\dots(5)$$

so when p is large

$$\phi(p) \simeq \frac{2}{\pi} \log\left(\frac{1 + p}{p}\right) = \frac{2}{\pi} \log(1 + 1/p) \simeq 2/\pi p. \quad \dots\dots(6)$$

Hence when t is small and positive

$$H_0(t) \simeq 2t/\pi. \quad \dots\dots\dots\dots\dots(7)$$

In this case if $p \rightarrow +\infty$, $H_0(t) \rightarrow 0$, which is the value at the origin. For the graph of $H_0(t)$ see Fig. 17, reference 11, p. 75.

2°. Sometimes $f(t)$ may be represented by

$$\sum_{r=0}^{\infty} a_r t^r + \log t \sum_{r=0}^{\infty} b_r t^r,$$

e.g. the Bessel functions $Y_0(t)$ and $K_0(t)$. Then since $\log t \rightarrow -\infty$ as $t \rightarrow +0$, the dominant term will have $\log t$ as a factor, and p can be finite, but large enough for the bulk of the Laplace integral to be contributed when t is small.

Example. What is the form of $Y_0(t)$ when t is small > 0?

$$Y_0(t) \rightleftharpoons -\frac{2p}{\pi\sqrt{p^2 + 1}} \log(p + \sqrt{p^2 + 1}), \quad \dots\dots(8)$$

so when p is large

$$\phi(p) \simeq -\frac{2}{\pi} \log 2p. \quad \dots\dots\dots\dots\dots(9)$$

Inverting this by aid of the list, we obtain

$$Y_0(t) \simeq \frac{2}{\pi}[\log (\tfrac{1}{2}t) + \gamma], \quad \dots\dots\dots\dots(10)^*$$

when t is small and positive.   Hence as $t \to +0$, $Y_0(t) \to -\infty$.

* $\gamma = 0.5772\dots$ which is negligible compared with $|\log(\tfrac{1}{2}t)|$ when $t < 10^{-4}$.

III

SOLUTION OF ORDINARY LINEAR DIFFERENTIAL EQUATIONS WITH CONSTANT COEFFICIENTS

3·11. Example illustrating procedure.

Solve
$$\frac{d^2y}{dt^2} + a^2y = e^{-bt}, \quad \dots\dots\dots\dots\dots(1)$$

when the initial conditions correspond to quiescence, i.e. $y = y' = 0$ at $t = 0$, and $b \gtrless 0$.

If (1) is multiplied throughout by pe^{-pt} and integrated with respect to t from 0 to $+\infty$,* each term in the modified equation is the L.T. of that in the original. In this way the differential equation is transformed into an algebraic one. Taking $y = f(t)$, we apply § 2·14 with $f(0) = f'(0) = 0$. Thus by (9) § 2·14,

$$(p^2 + a^2)\phi = p\int_0^\infty e^{-(p+b)t}\,dt = p/(p+b), \quad \dots\dots\dots(2)$$

provided $e^{-p_0 t}y$, $e^{-p_0 t}y'$ and $e^{-p_0 t}y''$ are bounded and continuous in $t \geqslant 0$,† p real $> p_0$, whilst $(p+b) > 0$ to ensure convergence of the integral. From the transform equation (2) we obtain

$$\phi = p/(p+b)(p^2+a^2). \quad \dots\dots\dots\dots\dots(3)$$

Then by (1) § 1·11 we have for y, the integral equation

$$p\int_0^\infty e^{-pt}y\,dt = p/(p+b)(p^2+a^2). \quad \dots\dots\dots(4)$$

Its solution is obtained by resolving $\phi/p = 1/(p+b)(p^2+a^2)$ into partial fractions‡ and expressing§ each of them in terms of t,

---

* This procedure for solving linear differential equations, as shown in this Chapter, was first used in reference 1, and is known as the pe^{-pt} method.

† This can be checked after the solution has been found. As stated in the footnote to (3) § 2·14, y and, therefore, y', y'' may be finitely discontinuous *at* $t = 0$. It is convenient to regard $(p^2+a^2)\phi = p/(p+b)$ as the transform equation.

‡ See footnote to § 3·131.

§ This is known as inversion.

by aid of the list of forms. Thus we find that

$$\phi = \frac{1}{(a^2 + b^2)}\left[\frac{p}{p+b} + \frac{pb}{p^2 + a^2} - \frac{p^2}{p^2 + a^2}\right], \quad \ldots\ldots\ldots(5)$$

so
$$y = \frac{1}{(a^2 + b^2)}\left[e^{-bt} + \frac{b}{a}\sin at - \cos at\right]. \quad \ldots\ldots\ldots(6)$$

If (6) is substituted into (1), the equation is satisfied. Also $y = 0$, $y' = 0$ when $t = 0$, so the initial conditions are reproduced, whilst $e^{-p_0 t}y$, $e^{-p_0 t}y'$, and $e^{-p_0 t}y''$ are bounded and continuous in $t \geqslant 0$, $p_0 \geqslant 0$. Hence (6) is a solution of (1). By Lerch's theorem in §§ 1·14, 1·16, this solution, continuous in the range $t \geqslant 0$, is unique.

The above procedure may be epitomised as follows :

(a) In (1) write ϕ for y, p for d/dt, and replace the r.h.s. by its L.T.

(b) Solve the resulting transform equation for ϕ, thereby obtaining (3) ;

(c) Resolve ϕ/p into partial fractions ;

(d) Invert each partial fraction, *in the expression for ϕ*, by aid of the list. This may be regarded as the partial-fraction-*cum*-list method.

3·12. Non-zero initial conditions in § 3·11.
We now introduce initial conditions concomitant with an acoustical, electrical, or other system, symbolised by (1) § 3·11, in action at $t = 0$. Then $y(0) = y_0$, and $y'(0) = y_1$, y_0, y_1 being non-zero. Applying (7) § 2·14 to the l.h.s. of (1) § 3·11 leads to

$$\frac{d^2y}{dt^2} + a^2 y \rightleftharpoons p^2\phi - p^2 y_0 - p y_1 + a^2 \phi.$$

Replacing e^{-bt} by its L.T., the transform equation is

$$(p^2 + a^2)\phi - p^2 y_0 - p y_1 = p/(p+b), \quad \ldots\ldots\ldots(1)$$

so
$$\phi = p\int_0^\infty e^{-pt}y\, dt = \frac{p}{(p+b)(p^2+a^2)} + \frac{p^2 y_0}{(p^2+a^2)} + \frac{p y_1}{(p^2+a^2)}. \quad \ldots(2)$$

This integral equation may be split up into the following component parts :

(a) the first member of the r.h.s., the inverse of which is the complete solution for quiescence initially;

(b) the second member on the r.h.s., whose inverse is that part of the solution due to the initial displacement y_0;

(c) the third member on the r.h.s., whose inverse is that part of the solution due to the initial rate of change of displacement y_1.

These three components are independent, but may be super-imposed in virtue of equation (1) § 3·11 being linear. The above analysis has the advantage not only of incorporating the initial conditions prior to solution, but also of displaying the influence of each condition separately, after inversion in terms of t. By aid of the list, (2) yields the continuous function

$$y = \frac{1}{(a^2+b^2)}\left[e^{-bt} + \frac{b}{a}\sin at - \cos at \right] + y_0 \cos at + \frac{y_1}{a}\sin at. \quad ...(3)$$

The reader should confirm that (3) satisfies (1) § 3·11 and the initial conditions ; also that $e^{-p_0 t}y$, $e^{-p_0 t}y'$ and $e^{-p_0 t}y''$ are bounded and continuous in $t > 0$. Consequently by Lerch's theorem §§ 1·14, 1·16, the solution (3) is unique.

Equation (1) § 3·11 can also be solved when the r.h.s. is zero,* provided that the initial conditions are incorporated in the transform equation. We then obtain

$$(p^2+a^2)\phi = p^2 y_0 + p y_1, \quad ....(4)$$

so
$$\phi = \frac{p^2 y_0}{p^2+a^2} + \frac{p y_1}{p^2+a^2} = y_0 \cos at + \frac{y_1}{a}\sin at. \quad ......(5)$$

Since $y = y_0$, $y' = y_1$ when $t = 0$, it follows that :

(i) the terms not containing y_0 must vanish at $t = 0$,

(ii) the first derivatives of the terms not containing y_1 must vanish at $t = 0$.

3·13. Ordinary linear D.E. of order n, with constant coefficients. The equation to be solved is

$$a_0 \frac{d^n y}{dt^n} + a_1 \frac{d^{n-1}}{dt^{n-1}}y + ... + a_n y = \xi(t), \quad ....(1)$$

the L.T. of $\xi(t)$ being assumed to exist. If (1) were the symbolical representation of a dynamical, electrical, or other

---

* This corresponds to the behaviour of any physical system represented by (1) § 3·11 *after* the driving force has been removed. Since $y \neq 0$, it follows that either one or both of y_0, y_1 is not zero.

system, $\xi(t)$ would represent the driving force or its equivalent. The initial conditions are symbolised by

$$y(0) = y_0, \ y'(0) = y_1, \ \ldots \ y^{(n-1)}(0) = y_{n-1}.$$

As in § 3·11, for quiescence initially, we replace each term in (1) by its L.T. using (9) § 2·14. If $y(t) \Rightarrow \phi(p)$, since $y_0 = y_1 = \ldots y_{n-1} = 0$, we obtain

$$(a_0 p^n + a_1 p^{n-1} + \ldots + a_n)\phi = p \int_0^\infty e^{-pt} \xi(t) dt = \phi_1. \quad \ldots\ldots(2)$$

Writing ϕ_2 for the polynomial in p, we obtain the transform equation

$$\phi = \phi_1/\phi_2, \quad \ldots\ldots\ldots\ldots\ldots\ldots\ldots\ldots(3)$$

so

$$p \int_0^\infty e^{-pt} y(t) dt = \phi_1/\phi_2. \quad \ldots\ldots\ldots\ldots\ldots\ldots(4)$$

This integral equation is solved for y by expressing $\phi_1/p\phi_2$* in partial fractions and using the list of transforms, or as shown in § 3·19.

3·131. Initial conditions. Here we apply (8) § 2·14, so to the l.h.s. of (2) § 3·13 we must add the contributions arising from the Σ terms corresponding to those on the l.h.s. of (1) § 3·13. Commencing with the term in a_0, we have the contributions:

$$\left.\begin{array}{ll} a_0: & -a_0[py_{n-1} + p^2 y_{n-2} + \ldots\ldots\ldots\ldots\ldots\ldots + \ p^n y_0] \\ a_1: & -a_1[\qquad py_{n-2} + p^2 y_{n-3} + \ldots\ldots\ldots\ldots + p^{n-1} y_0] \\ a_2: & -a_2[\qquad\qquad py_{n-3} + p^2 y_{n-4} + \ldots + p^{n-2} y_0] \\ \cdot & \quad \cdot \quad\quad \cdot \quad\quad \cdot \quad\quad \cdot \quad\quad \cdot \quad\quad \cdot \quad\quad \cdot \\ \cdot & \quad \cdot \quad\quad \cdot \quad\quad \cdot \quad\quad \cdot \quad\quad \cdot \quad\quad \cdot \quad\quad \cdot \\ a_{n-1}: & \quad \cdot \quad\quad \cdot \quad\quad \cdot \quad\quad \cdot \quad\quad \cdot \quad\quad -a_{n-1} p y_0. \end{array}\right\} \ldots(1)$$

The sum of these polynomials may be written in the form

$$-\left\{ a_0 \sum_{r=0}^{n-1} p^{n-r} y_r + a_1 \sum_{r=0}^{n-2} p^{n-r-1} y_r \right.$$

$$\left. + \ldots + a_{n-1} p y_0 \right\} = -\phi_3(p). \quad \ldots\ldots\ldots(2)$$

* See footnote to § 3·131.

We shall call $\phi_3(p)$ the L.T. of the initial conditions function. When it is incorporated in (2) § 3·13, we get the transform equation

$$\phi_2\phi = \phi_1 + \phi_3, \quad \dots\dots\dots\dots\dots(3)$$

or

$$p \int_0^\infty e^{-pt} y(t) dt = \phi = \frac{\phi_1}{\phi_2} + \frac{\phi_3}{\phi_2}, \quad \dots\dots\dots(4)$$

which leads to the integral equation for $y(t)$, namely,

$$\int_0^\infty e^{-pt} y(t) dt = \frac{\phi_1}{p\phi_2} + \frac{\phi_3}{p\phi_2}. \quad \dots\dots\dots(5)$$

$\phi_1/p\phi_2$ is now resolved into partial fractions * and the resulting expression for ϕ_1/ϕ_2 inverted using the list of forms. This gives the complete solution of (1) § 3·13 when the system is quiescent initially. Treating $\phi_3/p\phi_2$ in like manner yields that part of the solution of (1) § 3·13 arising from the initial conditions. Since the equation is a linear one, its two parts are independent but additive, i.e. that part of the solution arising from the initial conditions is unaffected by the form of $\xi(t)$. The contribution due to each condition may be segregated by selecting the appropriate column of terms in (1). Thus the L.T. of the contribution due to $y^{(n-r)}(0) = y_{n-r}$ is

$$y_{n-r} \sum_{m=1}^{r} a_{r-m} p^m / \phi_2(p). \quad \dots\dots\dots(6)$$

This is now to be inverted to obtain the desired result. A proof that the above procedure yields the correct result is given in Appendix V.

★3·132. Solution of (1) § 3·13 if $y^{(n)}(t) = y_n$, when $t = t_0$.

Let $t = \tau + t_0$, then

$$a_0 \frac{d^n y}{d\tau^n} + a_1 \frac{d^{n-1} y}{d\tau^{n-1}} + \dots + a_n y = \xi(\tau + t_0) \quad \dots\dots\dots(1)$$

and if $t = t_0$, $y_{\tau=0}^{(n)}(t) = y_n$. Thus replacing each term on the l.h.s. of (1) by its L.T., now defined as $\phi = p \int_0^\infty e^{-p\tau} y \, d\tau$, and

---

* It is preferable to resolve $\phi_1/p\phi_2$ rather than ϕ_1/ϕ_2 into partial fractions, each of which is multiplied by p before using the list for inversion.

adding the L.T. of the initial conditions function to the r.h.s., we get

$$(a_0 p^n + a_1 p^{n-1} + \ldots + a_n)\phi = p \int_0^\infty e^{-p\tau}\xi(\tau + t_0)\,d\tau + \phi_3. \quad \ldots\ldots(2)$$

Taking $p \int_0^\infty e^{-p\tau}\xi(\tau + t_0)\,d\tau = \bar{\phi}_1$, ϕ_2 as in § 3·13, and ϕ_3 as in § 3·131 gives

$$\phi = \frac{\bar{\phi}_1}{\phi_2} + \frac{\phi_3}{\phi_2}. \quad \ldots\ldots\ldots\ldots\ldots\ldots\ldots\ldots(3)$$

Thus by shifting the origin, (4) § 3·131 is obtained, except that ϕ_1 is replaced by $\bar{\phi}_1$. After the r.h.s. of (3) has been inverted in terms of τ, we write $\tau = (t - t_0)$.

3·14. Array for $\phi_3(p)$.

The polynomial for $\phi_3(p)$ is easily written down by using the Σ term in (8) § 2·14. Alternatively it may be expressed in the compact form of an array as follows :

$$\phi_3(p) = \begin{array}{c} \\ y_0 \\ y_1 \\ y_2 \\ \\ \\ y_{n-1} \end{array}
\begin{array}{c} p^n \quad\ p^{n-1} \quad\ p^{n-2} \ldots p \\
\left(\begin{array}{cccc}
a_0 & a_1 & a_2 & \ldots a_{n-1} \\
0 & a_0 & a_1 & \ldots a_{n-2} \\
0 & 0 & a_0 & \ldots a_{n-3} \\
\cdot & \cdot & \cdot & \ldots \\
\cdot & \cdot & \cdot & \ldots \\
0 & 0 & 0 & \ldots a_0
\end{array} \right)
\end{array} \quad \cdot \ \ldots\ldots(1)$$

Multiply each a by the corresponding y on its left, and then by the p immediately above. $\phi_3(p)$ is the sum of all such terms. Thus for an equation of the second order with $y' = y_1$, and $y = y_0$ at $t = 0$, we find that

$$\phi_3(p) = \begin{array}{c} \\ y_0 \\ y_1 \end{array} \begin{array}{c} p^2 \quad\ p \\ \left\{ \begin{array}{cc} a_0 & a_1 \\ 0 & a_0 \end{array} \right\} \end{array} = y_0 a_0 p^2 + (y_0 a_1 + y_1 a_0)p, \quad \ldots\ldots(2)$$

which is a useful formula to remember.

3·15. Practical procedure in solving (1) § 3·13.

1°. Obtain $\phi_1(p)$ the L.T. of $\xi(t)$ either by integration or from the list :

2°. Write p for d/dt on the l.h.s. thereby getting $\phi_2(p)$:

3°. Form $\phi_3(p)$ the L.T. of the initial conditions function using either the Σ term in (8) § 2·14, or (1) § 3·14:*

4°. Apply (4) § 3·131, resolve $(\phi_1/p\phi_2) + (\phi_3/p\phi_2)$ into partial fractions and use the list to interpret the resulting expressions for ϕ_1/ϕ_2, ϕ_3/ϕ_2; or apply the method of § 3·19. A proof that the procedure in §§ 3·13 *et seq.* gives the correct result will be found in Appendix V. By Lerch's theorem in §§ 1·14, 1·16 the solution is unique.

3·16. Example. Solve

$$\frac{d^2y}{dt^2} + a^2y = A \cos \omega t, \quad \omega \neq a;$$

1°. $y = y' = 0$ at $t = 0$; 2°. $y = y_0, y' = y_1$ at $t = 0$.

1°. Adopting the procedure in § 3·15, we have $\phi_3 = 0$, so the transform equation is

$$(p^2 + a^2)\phi = Ap^2/(p^2 + \omega^2), \quad \ldots\ldots\ldots\ldots\ldots(1)$$

or $\qquad y(t) \Rightarrow \phi(p) = Ap^2/(p^2 + \omega^2)(p^2 + a^2)$

$$= \frac{A}{(a^2 - \omega^2)}\left[\frac{p^2}{p^2 + \omega^2} - \frac{p^2}{p^2 + a^2}\right]. \quad \ldots\ldots\ldots(2)$$

From the list we find that

$$y = A (\cos \omega t - \cos at)/(a^2 - \omega^2). \quad \ldots\ldots\ldots\ldots(3)$$

2°. By (2) § 3·14 with $a_0 = 1$, $a_1 = 0$, the L.T. of the initial conditions function is

$$\phi_3(p) = y_0p^2 + y_1p, \quad \ldots\ldots\ldots\ldots\ldots\ldots(4)$$

so using (4) § 3·131, we obtain

$$y(t) \Rightarrow \phi(p) = \frac{Ap^2}{(p^2 + a^2)(p^2 + \omega^2)} + \frac{y_0p^2}{(p^2 + a^2)} + \frac{y_1p}{(p^2 + a^2)}. \quad \ldots\ldots(5)$$

In terms of t, we find from the list that

$$y = \frac{A}{(a^2 - \omega^2)}[\cos \omega t - \cos at] + y_0 \cos at + (y_1/a) \sin at. \quad \ldots(6)$$

The reader should check that (3), (6) satisfy the differential equation and the respective initial conditions.

* As an alternative to 2° and 3°, each term on the l.h.s. may be replaced by its L.T. using (8) § 2·14.

3·17. Example. Solve the equation in 1° § 3·16 if $a = \omega$.

Then $\qquad y(t) = \phi(p) = Ap^2/(p^2 + \omega^2)^2,$ ....(1)

so from the list, we find that the solution in terms of t is

$$y = \frac{At}{2\omega} \sin \omega t. \qquad \text{..................(2)}$$

The factor t is due to the repeated factor $(p^2 + \omega^2)^{-1}$ in (1), and the solution is non-periodic; $y \to \pm \infty$ as $t \to + \infty$.

*3·171. Example. Solve $\dfrac{d^2y}{dt^2} + a^2 y = \sin \omega t$, if $y = y_0$, $y' = y_1$, when $t = t_0$, $\omega \neq a$.

First we evaluate (see § 3·132)

$$\bar{\phi}_1 = p \int_0^\infty e^{-p\tau} \sin \omega (\tau + t_0) d\tau, \qquad \text{..................(1)}$$

$$= p \int_0^\infty e^{-p\tau} [\cos \omega t_0 \sin \omega \tau + \sin \omega t_0 \cos \omega \tau] d\tau,$$

$$= \frac{\omega p \cos \omega t_0}{p^2 + \omega^2} + \frac{p^2 \sin \omega t_0}{p^2 + \omega^2}. \qquad \text{..................(2)}$$

Then by (3) § 3·132 and (2) § 3·14, with $a_0 \doteq 1$, $a_1 = 0$,

$$\phi = \frac{p}{(p^2 + \omega^2)(p^2 + a^2)} [\omega \cos \omega t_0 + p \sin \omega t_0] + \frac{y_0 p^2}{p^2 + a^2} + \frac{y_1 p}{p^2 + a^2}. \quad (3)$$

In terms of τ, we find from the list that

$$y(\tau + t_0) = \frac{1}{(a^2 - \omega^2)} \left[\omega \cos \omega t_0 \left(\frac{\sin \omega \tau}{\omega} - \frac{\sin a\tau}{a} \right) \right.$$

$$\left. + \sin \omega t_0 (\cos \omega \tau - \cos a\tau) \right] + y_0 \cos a\tau + y_1 \frac{\sin a\tau}{a}. \quad \text{...(4)}$$

Writing $\tau = t - t_0$, (4) becomes

$$y(t) = \frac{1}{(a^2 - \omega^2)} \left[\omega \cos \omega t_0 \left\{ \frac{\sin \omega(t - t_0)}{\omega} - \frac{\sin a(t - t_0)}{a} \right\} \right.$$

$$\left. + \sin \omega t_0 \{ \cos \omega (t - t_0) - \cos a (t - t_0) \} \right]$$

$$+ y_0 \cos a(t - t_0) + y_1 \frac{\sin a(t - t_0)}{a}. \quad \text{......(5)}$$

3·18. Example illustrating equation with integral on l.h.s.

A potential difference $E = E_0 e^{-at}$ is applied to the series combination of inductor L and capacitor C (see Fig. 7), by closing the switch S at $t = 0$. As $t \to -0$ there is no current in the circuit, but C has a charge Q_0. What is the current at any time $t > 0$?

The differential equation for the current in the circuit is

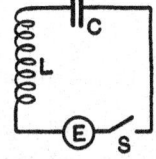

FIG. 7.—Schematic diagram of simple non-dissipative oscillatory electrical circuit.

$$L \frac{dI}{dt} + \frac{Q}{C} = E_0 e^{-at}, \quad \ldots\ldots\ldots\ldots(1)$$

where Q is the charge on the capacitor at any time $t > 0$. This is the sum of the initial charge Q_0 and that acquired in the interval 0 to t. Thus (1) may be expressed in the form

$$L \frac{dI}{dt} + \frac{1}{C} \int_0^t I \, dt + \frac{Q_0^*}{C} = E_0 e^{-at}. \quad \ldots\ldots\ldots\ldots(2)$$

There are two methods of solution.

Method 1. The *integral* in (2) being zero when $t = 0$ yields no initial conditions term, while that associated with $L dI/dt$ is also zero, since $I(0) = 0$ by hypothesis. Writing p for d/dt, p^{-1} for $\int_0^t dt$ (see § 2·15), and replacing the r.h.s. of (2) by its L.T., we obtain the transform equation

$$\left(Lp + \frac{1}{Cp} \right) \phi = \frac{E_0 p}{p+a} - \frac{Q_0}{C}, \quad \ldots\ldots\ldots\ldots(3)$$

where $I \rightleftharpoons \phi$.

Substituting $\omega^2 = 1/LC$, and rearranging terms, (3) becomes

$$\phi = \frac{(E_0/L)p^2}{(p+a)(p^2+\omega^2)} - \frac{\omega^2 p Q_0}{(p^2+\omega^2)}. \quad \ldots\ldots\ldots\ldots(4)$$

From the list, the inverse of (4) is

$$I = \frac{(E_0/L)}{(a^2+\omega^2)} \left[\omega \sin \omega t + a \cos \omega t - a e^{-at} \right] - \omega Q_0 \sin \omega t. \quad \ldots(5)$$

---

* The ' sense ' of the initial charge governs the sign of this term. We shall assume it to be positive.

Method 2. Since $I = \dfrac{dQ}{dt}$, the circuital equation (1) may be written

$$L\frac{d^2Q}{dt^2} + \frac{Q}{C} = E_0 e^{-at}. \qquad\qquad\ldots\ldots\ldots\ldots(6)$$

Using the initial conditions $Q = Q_0$, $\dfrac{dQ}{dt} = I = 0$ at $t = 0$, in (7) § 2·14, writing p for d/dt, taking $Q \Rightarrow \phi_1$, and replacing the r.h.s. by its transform, we obtain

$$\left(Lp^2 + \frac{1}{C}\right)\phi_1 = \frac{E_0 p}{p + a} + Lp^2 Q_0, \qquad\ldots\ldots\ldots\ldots(7)$$

or

$$\phi_1 = \frac{(E_0/L)p}{(p + a)(p^2 + \omega^2)} + \frac{Q_0 p^2}{(p^2 + \omega^2)}. \qquad\ldots\ldots\ldots(8)$$

By (5) § 2·14, with Q for $f(t)$, we have

$$I = \frac{dQ}{dt} \Rightarrow p[\phi_1 - Q(0)] = p[\phi_1 - Q_0], \qquad\ldots\ldots\ldots(9)$$

so by (8), (9)

$$I \Rightarrow p\left[\frac{(E_0/L)p}{(p + a)(p^2 + \omega^2)} + Q_0\left(\frac{p^2}{p^2 + \omega^2} - 1\right)\right], \qquad\ldots\ldots\ldots(10)$$

which is identical with (4), as we should expect. Q may also be found by inverting (8), and differentiating $Q(t)$ with respect to t. The reader should verify that (5) satisfies (1) and the initial conditions.

Mechanical analogue. A mass m, on a frictionless horizontal plane, is attached to one end of a massless helical spring of stiffness s, whose other end is fixed, the axis being horizontal. As $t \to -0$ the spring is compressed by an amount x_0, the corresponding force being x_0/c, where the compliance $c = 1/s$. An axial force $F_0 e^{-at}$ is applied to m at $t = 0$, so the appropriate differential equation is

$$m\frac{dv}{dt} + \frac{1}{c}\int_0^t v\,dt + \frac{x_0}{c} = F_0 e^{-at}, \qquad\ldots\ldots\ldots\ldots(11)$$

or

$$m\frac{d^2x}{dt^2} + \frac{x}{c} = F_0 e^{-at}. \qquad\qquad\ldots\ldots\ldots\ldots\ldots(12)$$

Equations (11) and (12) correspond to (2), (6) respectively, so m is analogous to L, $c = 1/s$ to C, v to I, x to Q, and F_0 to E_0, where v, x are axial velocity and displacement.

3·181. Electrical network symbolism. The potential differences across an inductance L, capacitance C, and resistance R due to a varying current I are, respectively, $L\,dI/dt, \dfrac{1}{C}\displaystyle\int_0^t I\,dt, RI,$ whilst the e.m.f. induced by virtue of mutual inductance M between two coils is $M\,dI/dt$. If $I \Rrightarrow \phi$, the L.T.S., for quiescence at $t=0$, are $pL\phi, \phi/pC, R\phi, pM\phi$, so that $pL, 1/pC, R, pM$ may be regarded as the transform impedances. Their reciprocals are the transform admittances.

3·182. Example illustrating § 3·181. If a p.d. $E = E_0 \cos \omega t$ is applied at $t=0$ to the circuit of Fig. 8, what is the current I at any time $t > 0$? E_0 is a constant, and the circuit is quiescent initially.

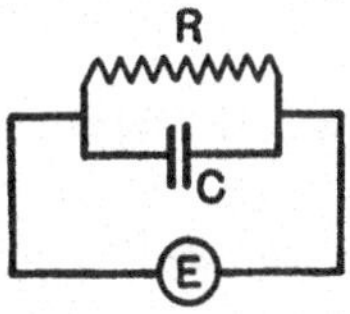

The transform admittance of R is $1/R$, and that of C is pC. Thus the total transform admittance is

$$\frac{1}{R} + pC = 1/Z(p), \quad \ldots\ldots\ldots\ldots(1)$$

where $Z(p)$ is the transform impedance.

Fig. 8.—Schematic diagram of simple electrical network.

If $I \Rrightarrow \phi, E \Rrightarrow \phi_1$, then since current = p.d./impedance, from preceding sections it follows that

$$\phi = \frac{\phi_1}{Z(p)} = \frac{E_0 p^2}{(p^2 + \omega^2)} \left(\frac{1}{R} + pC \right), \quad \ldots\ldots\ldots\ldots(2)$$

$$= (E_0/R) \frac{p^2}{(p^2 + \omega^2)} + E_0 Cp - \frac{E_0 C\omega^2 p}{(p^2 + \omega^2)}. \quad \ldots\ldots(3)$$

By the list we obtain

$$I = (E_0/R) \cos \omega t + E_0 C I(t) - E_0 C\omega \sin \omega t. \quad \ldots\ldots\ldots(4)$$

The first term on the r.h.s. of (4) represents the current through R, the second represents a current impulse of strength $E_0 C$, but of infinite amplitude, due to C acting momentarily as a short circuit, i.e. C is charged instantaneously, whilst the third gives the steady alternating current through C.

If C were replaced by an inductance L, we should have $1/Z(p) = \dfrac{1}{R} + \dfrac{1}{pL}$, so

$$\phi = \frac{E_0 p^2}{(p^2 + \omega^2)} \left(\frac{1}{R} + \frac{1}{Lp} \right) = (E_0/R) \frac{p^2}{(p^2 + \omega^2)} + (E_0/L) \frac{p}{(p^2 + \omega^2)}, \quad (5)$$

and by the list

$$I = (E_0/R) \cos \omega t + (E_0/\omega L) \sin \omega t. \quad \dots\dots\dots(6)$$

The first term on the r.h.s. of (6) represents the current in R, and the second term that in L.

*3·19.   Application of product theorem to find solution of (1) § 3·13 for initial quiescence.

1°. By (4) § 3·131 we have to solve the integral equation

$$y_p(t) \Rightarrow \phi_1/\phi_2, \quad \dots\dots\dots\dots\dots(1)$$

where $\xi(t) \Rightarrow \phi_1$. If $\chi(t) \Rightarrow p/\phi_2$, then by § 2·24

$$y_p(t) = \int_0^t \chi(\lambda)\xi(t-\lambda)d\lambda = \int_0^t \chi(t-\lambda)\xi(\lambda)d\lambda. \quad \dots\dots(2)$$

2°. If $\zeta(t) \Rightarrow 1/\phi_2$, then by § 2·242

$$y_p(t) = \frac{d}{dt}\int_0^t \zeta(\lambda)\xi(t-\lambda)d\lambda = \frac{d}{dt}\int_0^t \zeta(t-\lambda)\xi(\lambda)d\lambda. \quad \dots\dots(3)$$

Using (1)–(4) § 2·243, the integrals in (3) may be expressed in an alternative form.

When $\phi_1 = 1$ in (1), $\xi(t) = H(t)$ and $y_p(t) = \zeta(t)$. Hence when the solution $\zeta(t)$ for the Heaviside unit function is known, that for *any* function $\xi(t)$, which has a L.T., is found from (3).

*3·191. As a matter of interest we shall apply (5) § 2·14 to (2) § 3·19 to obtain an alternative form of (3) § 3·19. Since $\zeta(t) \Rightarrow 1/\phi_2$, by (5) § 2·14

$$\zeta'(t) \Rightarrow p/\phi_2 - p\zeta(0). \quad \dots\dots\dots\dots(1)$$

Now $\chi(t) \Rightarrow p/\phi_2$, so by (1)

$$\chi(t) = \zeta'(t) + \zeta(0)I(t), \quad \dots\dots\dots\dots(2)$$

where $I(t)$ is the impulsive function of § 2·19.   Applying (2) to (2) § 3·19, we get

$$y_p(t) = \int_0^t [\zeta'(\lambda) + \zeta(0)I(\lambda)]\xi(t-\lambda)d\lambda,$$

$$= \zeta(0)\xi(t) + \int_0^t \zeta'(\lambda)\xi(t-\lambda)d\lambda, \quad \dots\dots\dots(3)$$

by (7) § 2·191, provided $\zeta(t)$ and $\xi(t)$ are continuous.   Formula (3) is identical in form with (2) § 2·243.

3·21. Simultaneous ordinary linear differential equations of the first order with constant coefficients.

Let the equations for solution be

$$a_0 \frac{dy}{dt} + a_1 \frac{dx}{dt} + a_2 y = \xi_1(t), \quad \dots\dots\dots\dots(1)$$

$$b_0 \frac{dx}{dt} + b_1 \frac{dy}{dt} + b_2 x = \xi_2(t), \quad \dots\dots\dots\dots(2)$$

where x, y, ξ_1, ξ_2 are functions of t, and the initial conditions are $x = x_0$, $y = y_0$ at $t = 0$. Each term in (1), (2) is replaced by its L.T. using (5) § 2·14. Taking $x(t) \Rightarrow \phi_0(p)$, $y(t) \Rightarrow \phi(p)$, $\xi_1(t) \Rightarrow \phi_1(p)$, and $\xi_2(t) \Rightarrow \phi_2(p)$, we obtain the transform equations:

$$a_1 p \phi_0 + (a_0 p + a_2)\phi = \phi_1 + p(a_0 y_0 + a_1 x_0), \quad \dots\dots\dots(3)$$

$$(b_0 p + b_2)\phi_0 + b_1 p \phi = \phi_2 + p(b_0 x_0 + b_1 y_0). \quad \dots\dots\dots(4)$$

By solving (3), (4) as simultaneous algebraic equations, ϕ_0 and ϕ are obtained in terms of p and the other quantities. Let the transform solutions so found be

$$\phi_0 = \psi_0(p), \quad \dots\dots\dots\dots\dots\dots(5)$$

and

$$\phi = \psi(p). \quad \dots\dots\dots\dots\dots\dots(6)$$

The ultimate solutions are the inversions of ψ_0 and ψ in terms of t. They may be obtained (1) by reducing ψ_0 and ψ to partial fractions and using the list, (2) by the method used in § 3·19 (the product theorem), or (3) by the Mellin inversion theorem (see Appendix IV). Various conditions regarding convergence, etc., stipulated in §§ 2·14, 3·11, are assumed to be satisfied.

3·211. Example.

Referring to Fig. 9, a p.d. $E = E_0 H(t)$—see Fig. 13—is applied to the primary circuit of an air-cored transformer. If the system is quiescent initially, what is the current I_2 when $t > 0$?

Let $I_1 \Rightarrow \phi_1$, $I_2 \Rightarrow \phi_2$, then equating the transforms of the p.d.s. in each circuit to zero *

$$(R_1 + pL_1)\phi_1 + pM\phi_2 - E_0 = 0, \quad (1)$$

and $(R_2 + pL_2)\phi_2 + pM\phi_1 \quad = 0. \quad (2)$

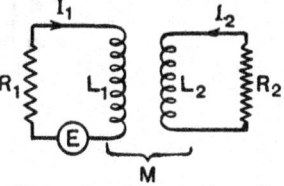

FIG. 9.—Schematic diagram of simple electrical air-cored transformer network.

* If $f(t)$ represents a p.d. or an e.m.f., by Kirchhoff's law $\sum\limits_{r=1}^{n} f_r(t) = 0$ round any closed circuit. Hence

$$p \int_0^{\infty} e^{-pt} \sum_{r=1}^{n} f_r(t)\, dt = \sum_{r=1}^{n} \phi_r(p) = 0.$$

Substituting for ϕ_1 from (2) into (1), we get

$$[(R_1 + pL_1)(R_2 + pL_2) - p^2 M^2]\phi_2 = -pME_0, \dotfill (3)$$

so $$\phi_2 = -pcE_0/(p^2 + ap + b), \dotfill (4)$$

where $$a = \left(\frac{R_1}{L_1} + \frac{R_2}{L_2}\right)\Big/(1 - k^2), \quad b = \left(\frac{R_1 R_2}{L_1 L_2}\right)\Big/(1 - k^2),$$

$$c = k/(1 - k^2)\sqrt{L_1 L_2}, \qquad k = M/\sqrt{L_1 L_2}.$$

Now (4) may be written

$$\phi_2 = -cE_0 p/[(p + \tfrac{1}{2}a)^2 - (\tfrac{1}{4}a^2 - b)], \quad \dotfill (5)$$

so by the list and § 2·13

$$I_2 = -\frac{cE_0 e^{-\frac{1}{2}at}}{\alpha}\sinh \alpha t, \quad \dotfill (6)$$

where $\alpha = \sqrt{\tfrac{1}{4}a^2 - b}$, $\tfrac{1}{4}a^2 > b$. The negative sign indicates reversal of phase due to e.m. induction. The current attains a maximum value and sinks asymptotically to zero thereafter.

Solution when the p.d. is $E = f_0(t) \rightleftharpoons \phi_0(p)$. (5) now takes the form

$$\phi_2 = -cp\phi_0/[(p + \tfrac{1}{2}a)^2 - (\tfrac{1}{4}a^2 - b)] \dotfill (7)$$

$$= -c \cdot \phi\phi_0, \dotfill (8)$$

where $$\phi = p/[(p + \tfrac{1}{2}a)^2 - (\tfrac{1}{4}a^2 - b)] \rightleftharpoons \frac{e^{-\frac{1}{2}at}}{\alpha}\sinh \alpha t.$$

Hence by § 2·243

$$I_2 = -\frac{c}{\alpha}\frac{d}{dt}\int_0^t f_0(t - \lambda)e^{-\frac{1}{2}a\lambda}\sinh \alpha\lambda \, d\lambda, \quad \dotfill (9)$$

$$= -\frac{c}{\alpha}\frac{d}{dt}\int_0^t f_0(\lambda)e^{-\frac{1}{2}a(t-\lambda)}\sinh \alpha(t - \lambda)d\lambda. \quad \dotfill (10)$$

IV

SOLUTION OF PARTIAL LINEAR DIFFERENTIAL EQUATIONS WITH CONSTANT COEFFICIENTS

4·11. The method of solution is similar to that used in Chapter III, and is exemplified in the elementary case treated below. Solve the equation

$$\frac{\partial^2\theta}{\partial x^2} - \frac{1}{k}\frac{\partial\theta}{\partial t} = 0, \quad \dots\dots\dots\dots\dots(1)$$

subject to the initial condition :

$\theta(x, t) = \theta_0$ a constant, $t \to 0$, $x > 0$, i.e. $0 < x < +\infty$;

and the boundary conditions :

(a) $\theta(x, t) = \theta_0$ a constant, $x \to +\infty$, $t > 0$, i.e. $0 < t < +\infty$;

(b) $\theta(x, t) = \theta_1$ a constant, $x \to 0$, $t > 0$, i.e. $0 < t < +\infty$.

The above equation occurs in a number of technical applications, such as the diffusion of heat in a homogeneous bar of uniform cross-section whose lateral surface is thermally insulated, diffusion of water vapour through walls, and transmission of electrical energy in a uniform submarine telegraph cable whose inductance and leakance are negligible. In heat diffusion and electrical transmission, $\theta(x, t)$ represents, respectively, temperature difference and potential difference between a fixed datum and a point distant x from the origin at time t. k is the coefficient of diffusion in heat problems = thermal conductivity/(density × specific heat).

In the case of a bar, the conditions specified above assert that prior to $t = 0$ the whole is at a temperature θ_0. At $t = 0$, the temperature at the free end $x = 0$ is raised to θ_1 and maintained at that value as t increases.

To solve (1), we commence by replacing each member by its L.T., taking $\phi(x, p) = p \int_0^\infty e^{-pt}\theta(x, t)dt$. Using the second formula in the footnote to (8) § 2·14, $\frac{\partial\theta}{\partial t} \rightleftharpoons p(\phi - \theta_0)$, so the transform equation for (1) is

$$p\int_0^\infty e^{-pt}\left(\frac{\partial^2\theta}{\partial x^2}\right)dt - \frac{p\phi}{k} = -\frac{p\theta_0}{k}. \quad \dots\dots\dots(2)$$

We now *assume* that the order of differentiation and integration in (2) may be inverted, i.e.

$$p \int_0^\infty e^{-pt} \left(\frac{\partial^2 \theta}{\partial x^2} \right) dt = p \frac{\partial^2}{\partial x^2} \int_0^\infty e^{-pt} \theta(x,t) \, dt = \frac{d^2 \phi}{dx^2}. \quad \ldots\ldots(3)$$

The validity of this step depends upon the nature of the function $\theta(x, t)$ and will be investigated later. By (2) and (3), the transform equation may be written

$$\frac{d^2 \phi}{dx^2} - \frac{p\phi}{k} = -\frac{p\theta_0}{k}. \quad \ldots\ldots\ldots\ldots\ldots(4)$$

In this way a partial differential equation may be transformed into an ordinary differential equation. The formal solution of (4) is

$$\phi = A e^{-x\sqrt{p/k}} + B e^{x\sqrt{p/k}} + \theta_0, \quad \ldots\ldots\ldots\ldots(5)$$

where A and B are arbitrary constants, or functions of p, which depend upon the boundary conditions. Since

$$\phi = p \int_0^\infty e^{-pt} \theta(x, t) \, dt, \quad \ldots\ldots\ldots\ldots(6)$$

we have (a) $\phi = \theta_0$, $x \to \infty$, $t > 0$; (b) $\phi = \theta_1$, $x = 0$, $t > 0$.

Applying (a) to (5) leads to $B = 0$, since both ϕ and θ must be finite, whilst (b) entails $A = (\theta_1 - \theta_0)$. Hence

$$\phi = \theta_0 + (\theta_1 - \theta_0) e^{-x\sqrt{p/k}} = p \int_0^\infty e^{-pt} \theta(x, t) \, dt, \quad \ldots\ldots\ldots(7)$$

so this has to be solved as an integral equation for $\theta(x, t)$. By aid of the list we find the solution to be

$$\theta(x, t) = \theta_0 + (\theta_1 - \theta_0) \operatorname{erfc}(x/2\sqrt{kt}). \quad \ldots\ldots\ldots(8)$$

The graphs of erf u and erfc u are given in Fig. 10.

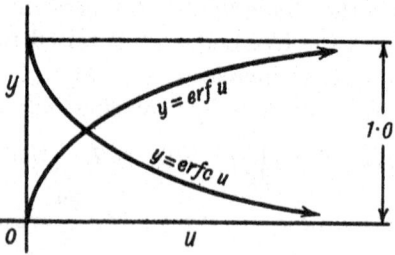

FIG. 10.—Graphs illustrating the forms of the monotonic functions erf u, erfc $u = 1 - \operatorname{erf} u$ for $u \geqslant 0$. erf$(-u) = -\operatorname{erf} u$, erfc$(-u) = 1 + \operatorname{erf} u$, so erfc$(-u) \pm \operatorname{erfc} u = 2$ or $2 \operatorname{erf} u$.

4·111. Proof that (8) § 4·11 is the appropriate solution of (1) § 4·11.

It is not possible to prove that in general the procedure for solving P.D.E., used in § 4·11, leads to the correct result. Accordingly each solution must be verified independently. The verification is divisible into four parts :

1°. To show that (8) § 4·11 satisfies the D.E. (1) § 4·11 ;
2°.   ,,   ,,   ,,   ,,   initial condition(s) ;
3°.   ,,   ,,   ,,   ,,   boundary   ,,
4°. To show that $e^{-p_0 t}\theta$, $e^{-p_0 t}\partial\theta/\partial x$, and $e^{-p_0 t}\partial^2\theta/\partial x^2$ are bounded and continuous in $t\geqslant 0$, $x>0$, $p_0>0$.

For the sake of completeness, the validity of the inverted order of differentiation and integration in (3) § 4·11 may be investigated.

1°. Let
$$v=\operatorname{erfc} u=\frac{2}{\sqrt{\pi}}\int_u^\infty e^{-y^2}\,dy, \quad\ldots\ldots\ldots\ldots\ldots(1)$$

where $u=x/2\sqrt{kt}$. Then
$$\frac{\partial v}{\partial x}=\frac{\partial v}{\partial u}\cdot\frac{\partial u}{\partial x}=-\frac{2}{\sqrt{\pi}}e^{-x^2/4kt}\cdot\frac{1}{2\sqrt{kt}}=-e^{-x^2/4kt}/\sqrt{\pi kt}, \quad\ldots\ldots(2)$$

and
$$\frac{\partial^2 v}{\partial x^2}=\frac{x}{2\sqrt{\pi}}\cdot\frac{1}{(kt)^{3/2}}\cdot e^{-x^2/4kt}. \quad\ldots\ldots\ldots\ldots\ldots(3)$$

Also
$$\frac{\partial v}{\partial t}=\frac{\partial v}{\partial u}\cdot\frac{\partial u}{\partial t}=-\frac{2}{\sqrt{\pi}}e^{-x^2/4kt}(-x/4k^{1/2}t^{3/2})$$
$$=xe^{-x^2/4kt}/2\sqrt{\pi k}\,t^{3/2}. \quad\ldots\ldots\ldots\ldots\ldots(4)$$

Hence by (3) and (4)
$$\frac{\partial^2 v}{\partial x^2}-\frac{1}{k}\frac{\partial v}{\partial t}=0, \quad\ldots\ldots\ldots\ldots\ldots(5)$$

so (8) § 4·11 satisfies (1) § 4·11.

2°. We take the initial condition in the form $\lim \theta(x,t)\to\theta_0$ as $t\to+0$ for any fixed positive value of x. This is satisfied by (8) § 4·11 when $x>0$, since erfc $x/2\sqrt{kt}\to 0$ as $t\to+0$ (see Fig. 10).

3°. Taking the first boundary condition, erfc $u\to 0$ as $u\to+\infty$ for any fixed $t>0$, so $\theta=\theta_0$ as required. For the second boundary condition $\lim \theta(x, t) \to \theta_1$, $x \to +0$, for any fixed $t > 0$, since erfc $u \to 1$, so this condition is satisfied also.

$4°$. Finally, $e^{-p_0 t}\theta$ in (8) § 4·11 is bounded and continuous in $t \geqslant 0$, $x \geqslant 0$, $p_0 \geqslant 0$, whilst $e^{-p_0 t}(\partial\theta/\partial x)$ and $e^{-p_0 t}(\partial^2\theta/\partial x^2)$ are b. and c. in $t \geqslant 0$, $x > 0$, $p_0 > 0$ (see (2) above and § 4·213).

4·112. Proof that (3) § 4·11 is valid. By (8) § 4·11 and (3) § 4·111

$$\frac{1}{(\theta_1 - \theta_0)} \frac{\partial^2\theta}{\partial x^2} = \frac{x}{2\sqrt{\pi}} (kt)^{-3/2} e^{-x^2/4kt}, \quad \ldots\ldots\ldots\ldots(1)$$

so by the list

$$\frac{1}{(\theta_1 - \theta_0)} p \int_0^\infty e^{-pt} \left(\frac{\partial^2\theta}{\partial x^2}\right) dt = \frac{p}{k} e^{-x\sqrt{p/k}}. \quad \ldots\ldots\ldots\ldots(2)$$

Also by (3), (7) § 4·11

$$\frac{1}{(\theta - \theta_0)} p \frac{\partial^2}{\partial x^2} \int_0^\infty e^{-pt} \theta(x, t) dt = \frac{p}{k} e^{-x\sqrt{p/k}}. \quad \ldots\ldots\ldots\ldots(3)$$

Since (2), (3) are identical it follows that inversion of the order of differentiation and integration in (3) § 4·11 is permissible. This completes the proof that (8) § 4·11 is a solution of (1) § 4·11 which satisfies the given initial and boundary conditions. In virtue of Lerch's theorem in §§ 1·14, 1·16, it is unique.

4·12. Procedure in § 4·11. This may be epitomised in the following way : In (1) § 4·11 substitute ϕ for θ, p for $\partial/\partial t$, and add the initial conditions function using the Σ term in (8) § 2·14. Alternatively, replace each member of (1) § 4·11 by its L.T. using the formula in the footnote to § 2·14. Solve this transform equation (which is an ordinary differential equation in x) formally for ϕ, and insert the boundary conditions, thereby ascertaining the arbitrary constants. ϕ is a function of (p, x) which is now inverted in terms of t by aid of the list. The inverse $\theta(x, t)$ is the solution of the integral equation

$$p \int_0^\infty e^{-pt}\theta(x, t) dt = \phi(p, x). \quad \ldots\ldots\ldots\ldots\ldots(1)$$

In some problems the inversion may present considerable difficulty. It often happens, however, that the solution of (1) is given by the complex integral part of the Mellin inversion theorem, as stated in Appendix IV (see also reference 12). Having obtained a solution, its appropriateness should be checked using the procedure in §§ 4·111, 4·112.

4·13. Initial condition in (1) § 4·11 an arbitrary function of x. The temperature distribution in a very long thermally insulated homogeneous metal bar of uniform cross-section is maintained at $\theta = \theta_0 e^{-ax}$ by an external agency until $t = 0$. When $t > 0$, the end $x = 0$ is kept at θ_0, whilst heat diffuses down the bar due to : (a) the supply at temperature θ_0, (b) the redistribution following relaxation of the control at $t = 0$. Assuming that the bar is long enough for the effect of reflection from the far end, at a point sufficiently remote therefrom, to be negligible, what is the temperature distribution along the bar at any time $t > 0$?

We have to solve (1) § 4·11 for

1° the initial condition $\qquad \theta(x, t) = \theta_0 e^{-ax}$, $t \to 0$, $x > 0$;

2° the boundary condition $\theta(x, t) \to 0$, $x \to +\infty$, $t > 0$;

3° the boundary condition $\theta(x, t) = \theta_0$, a constant, $x \to 0$, $t > 0$.

Proceeding as in § 4·11 and making the necessary assumption respecting inversion of the order of differentiation and integration of the integrals, we obtain the transform equation

$$\frac{d^2\phi}{dx^2} - \frac{p\phi}{k} = -\frac{p\theta_0 e^{-ax}}{k} , \qquad \ldots\ldots\ldots\ldots\ldots(1)$$

of which the solution is

$$\phi = A e^{-x\sqrt{p/k}} + B e^{x\sqrt{p/k}} + \frac{p\theta_0 e^{-ax}}{p - a^2 k} . \qquad \ldots\ldots\ldots\ldots(2)$$

Inserting the first boundary condition in (2) leads to $B = 0$, since ϕ and θ must tend to zero as $x \to +\infty$. For the second boundary condition $\phi = \theta_0$ at $x = 0$, we get

$$A = -\frac{\theta_0 a^2 k}{(p - a^2 k)} . \qquad \ldots\ldots\ldots\ldots\ldots\ldots(3)$$

---

* For a short bar the influence of reflection is important and cannot be neglected at intermediate points. In a mathematical sense, a very long bar may be taken as $\to +\infty$.

Substituting for A and B into (2) gives

$$\phi = \theta_0 \left[e^{-ax} \frac{p}{(p-c)} - c \frac{e^{-x\sqrt{p/k}}}{(p-c)} \right], \qquad \ldots\ldots\ldots\ldots(4)$$

where $c = a^2 k$. By aid of the list and (4) § 2·11, we find that

$$\theta = \theta_0 \left[\operatorname{erfc}(x/2\sqrt{kt}) + e^{a^2 kt} \times \right.$$

$$\left. \times \left\{ e^{-ax} - \tfrac{1}{2}e^{-ax} \operatorname{erfc}\left(\frac{x-2kat}{2\sqrt{kt}} \right) - \tfrac{1}{2}e^{ax} \operatorname{erfc}\left(\frac{x+2kat}{2\sqrt{kt}} \right) \right\} \right]. \quad (5)$$

Validity checking similar to that in §§ 4·111, 4·112 is left as a suitable exercise for the reader.

*4·14. Example. Solve $\dfrac{\partial^2 y}{\partial x^2} + a \dfrac{\partial y}{\partial x} = b^2 \dfrac{\partial y}{\partial t}$, subject to quiescence initially, and to the boundary conditions: (1) $y \to 0$ when $x \to +\infty$, (2) $\dfrac{\partial y}{\partial x} = y_1$ a constant, when $x = 0$. a, b are real and >0.

Hitherto we have confined our attention to P.L.D.E. having only the second derivative on the l.h.s. In the present instance the first derivative in x is included also, and this results in a somewhat complicated solution involving error function integrals.

To obtain the transform equation we write $y(x, t) \rightleftharpoons \phi(x, p)$, which gives

$$\frac{d^2\phi}{dx^2} + a\frac{d\phi}{dx} - b^2 p\phi = 0. \qquad \ldots\ldots\ldots\ldots\ldots(1)$$

The formal solution of (1) is

$$\phi = A e^{\lambda_1 x} + B e^{\lambda_2 x}, \qquad \ldots\ldots\ldots\ldots\ldots(2)$$

where λ_1, λ_2 are the roots of $\lambda^2 + a\lambda - b^2 p = 0$. Thus

$$\left.\begin{array}{c}\lambda_1 \\ \lambda_2\end{array}\right\} = -\tfrac{1}{2}[a \pm \sqrt{(a^2 + 4b^2 p)}] = -\tfrac{1}{2}[a \pm 2b\sqrt{(p+c^2)}], \quad \ldots(3)$$

with $c^2 = a^2/4b^2$. Since $4b^2 p > 0$, it follows that $\lambda_1 < 0$, $\lambda_2 > 0$. Now $y = 0$ and, therefore, $\phi = 0$ when $x = +\infty$, so with $\lambda_2 > 0$, B must be zero. Also

$$\left(\frac{\partial\phi}{\partial x} \right)_{x=0} = [A\lambda_1 e^{\lambda_1 x}]_{x=0} = A\lambda_1. \qquad \ldots\ldots\ldots\ldots(4)$$

By hypothesis

$$p \int_0^\infty e^{-pt} y \, dt = \phi, \quad \text{.............(5)}$$

so if differentiation of y, with respect to x, under the integral sign is permissible, we have

$$\frac{\partial \phi}{\partial x} = p \frac{\partial}{\partial x} \int_0^\infty e^{-pt} y \, dt = p \int_0^\infty e^{-pt} \left(\frac{\partial y}{\partial x} \right) dt. \quad \text{............(6)}$$

The validity of this step can be checked later. Since

$$p \int_0^\infty e^{-pt} \, dt = 1 \quad \text{and} \quad \left(\frac{\partial y}{\partial x} \right)_{x=0} = y_1, \text{ a constant,}$$

it follows from (6) that

$$y_1 = \left(\frac{\partial y}{\partial x} \right)_{x=0} = \left(\frac{\partial \phi}{\partial x} \right)_{x=0}. \quad \text{.................(7)}$$

Thus by (4) and (7),

$$A = y_1 / \lambda_1. \quad \text{.................(8)}$$

Substituting for A, B in (2) leads to the transform solution

$$\phi = y_1 e^{\lambda_1 x} / \lambda_1, \quad \text{.................(9)}$$

so we have now to solve the integral equation

$$p \int_0^\infty e^{-pt} y(x, t) \, dt = y_1 e^{\lambda_1 x} / \lambda_1. \quad \text{.................(10)}$$

By (6) and (10) above

$$p \int_0^\infty e^{-pt} \left(\frac{\partial y}{\partial x} \right) dt = y_1 e^{\lambda_1 x}, \quad \text{.................(11)}$$

with $\lambda_1 = -\frac{1}{2}[a + 2b(p + c^2)^{1/2}]$. From the list we find that

$$\frac{\partial y}{\partial x} = \frac{1}{2} y_1 (e^{-ax} \operatorname{erfc} u + \operatorname{erfc} v), \quad \text{.................(12)}$$

where

$$\left. \begin{array}{c} u \\ v \end{array} \right\} = \frac{b^2 x \mp at}{2b\sqrt{t}}.$$

When $x = 0$ and $a\sqrt{t}/2b = l$, (12) reduces to

$$\left(\frac{\partial y}{\partial x} \right)_{x=0} = \frac{1}{2} y_1 [\operatorname{erfc}(-l) + \operatorname{erfc}(l)] = y_1 \quad \text{..........(13)}$$

(see caption to Fig. 10) so the boundary condition is satisfied.

*4·15. Evaluation of y in (12) § 4·14.

We have

$$y = \tfrac{1}{2}y_1\left\{\int_0^x e^{-ax}\,\text{erfc}\,u\,dx + \int_0^x \text{erfc}\,v\,dx\right\} + \chi(t), \quad\ldots\ldots..(1)$$

where $\chi(t)$ is a function of t, independent of x.

1°. Consider the first integral in (1). By definition

$$\text{erfc}\,u = \frac{2}{\sqrt{\pi}}\int_u^\infty e^{-r^2}\,dr, \quad\ldots\ldots\ldots\ldots\ldots(2)$$

so

$$\frac{d\,\text{erfc}\,u}{dx} = -\frac{b}{\sqrt{\pi t}}e^{-u^2}. \quad\ldots\ldots\ldots\ldots\ldots(3)$$

Then

$$\int_0^x e^{-ax}\,\text{erfc}\,u\,dx = -\frac{1}{a}\int_0^x \text{erfc}\,u\,d(e^{-ax}), \quad\ldots\ldots\ldots\ldots(4)$$

$$= -\frac{1}{a}\left[e^{-ax}\,\text{erfc}\,u\right]_0^x + \frac{1}{a}\int_0^x e^{-ax}\left(\frac{d\,\text{erfc}\,u}{dx}\right)dx, \quad\ldots\ldots..(5)$$

$$= -\frac{1}{a}e^{-ax}\,\text{erfc}\,u + \frac{1}{a}\text{erfc}\,(-l) - \frac{b}{a\sqrt{\pi t}}\int_0^x e^{-ax-u^2}\,dx. \quad\ldots..(6)$$

Since $-(ax+u^2) = -(b^2x+at)^2/4b^2t = -v^2$, the integral on the r.h.s. of (6) may be written

$$\frac{b}{a\sqrt{\pi t}}\int_0^x e^{-v^2}\,dx = \frac{2}{a\sqrt{\pi}}\int_l^v e^{-v^2}\,dv = \frac{1}{a}[\text{erfc}\,l - \text{erfc}\,v]. \quad\ldots\ldots(7)$$

Hence by (6), (7)

$$\int_0^x e^{-ax}\,\text{erfc}\,u\,dx = -\frac{1}{a}[e^{-ax}\,\text{erfc}\,u - \text{erfc}\,v - 2\,\text{erf}\,l]. \quad\ldots..(8)$$

2°. Consider the second integral in (1). Here

$$\int_0^x \text{erfc}\,v\,dx = \left[x\,\text{erfc}\,v\right]_0^x - \int_0^x x\left(\frac{d}{dx}\text{erfc}\,v\right)dx, \quad\ldots\ldots..(9)$$

$$= x\,\text{erfc}\,v + \frac{b}{\sqrt{\pi t}}\int_0^x xe^{-v^2}\,dx. \quad\ldots\ldots\ldots..(10)$$

Writing $k = b/2\sqrt{t}$, we have

$$\int_0^x e^{-v^2}x\,dx = \frac{1}{2k^2}\int_0^x e^{-v^2}\,d(v^2) - \frac{l}{k}\int_0^x e^{-v^2}\,dx, \quad\ldots\ldots\ldots(11)$$

$$= \frac{1}{2k^2}(e^{-l^2} - e^{-v^2}) - \frac{\sqrt{\pi t}}{b}\frac{l}{k}[\text{erfc}\,l - \text{erfc}\,v], \quad\ldots\ldots(12)$$

by (7). Hence by (10) and (12)

$$\int_0^x \text{erfc } v \, dx = x \text{ erfc } v + \frac{l}{k} (\text{erfc } v - \text{erfc } l) + \frac{1}{k\sqrt{\pi}} (e^{-l^2} - e^{-v^2}). \quad (13)$$

From (1), (8), (13) we find that

$$y = \tfrac{1}{2} y_1 \left\{ \left(\frac{1}{a} + x + \frac{at}{b^2} \right) \text{erfc } v - \frac{e^{-ax} \text{erfc } u}{a} - \frac{2}{b} \sqrt{\frac{t}{\pi}} e^{-v^2} \right\}$$

$$+ \tfrac{1}{2} y_1 \left[\frac{2}{a} \text{erf } l - \frac{at}{b^2} \text{erfc } l + \frac{2}{b} \sqrt{\frac{t}{\pi}} e^{-l^2} \right] + \chi(t), \quad ...(14)$$

where $\chi(t)$ is a function of t independent of x.

For large values of w, $\text{erfc } w \sim e^{-w^2}/w\sqrt{\pi}$ [reference 12, p. 101], so when $x = +\infty$, the expression in { } vanishes. Since the first boundary condition necessitates $y \to 0$, $x \to +\infty$, it follows that

$$\chi(t) = -\tfrac{1}{2} y_1 [\quad]. \quad \quad \quad ...(15)$$

Hence in terms of t, the solution of (1) § 4·14 is

$$y = \tfrac{1}{2} y_1 \left\{ \left(\frac{1}{a} + x + \frac{at}{b^2} \right) \text{erfc } v - \frac{e^{-ax} \text{erfc } u}{a} - \frac{2}{b} \sqrt{\frac{t}{\pi}} e^{-v^2} \right\}. \quad ...(16)$$

We have shown already that (16) satisfies both of the boundary conditions. When $t \to +0$, so also does (16), and the initial state of quiescence is satisfied.

4·16. Plane Sound Waves in a Viscous Medium.

A rigid disk vibrates at one end of a uniform frictionless cylindrical tube of equal radius. The tube is either sufficiently long or terminated by an acoustical impedance to make reflection at the far end negligible, i.e. the transmission is unidirectional. The differential equation for sinusoidal sound waves of very small amplitude is

$$\rho_0 c^2 \frac{\partial^2 \xi}{\partial x^2} - \rho_0 \frac{\partial^2 \xi}{\partial t^2} - r \frac{\partial \xi}{\partial t} = 0, \quad \quad ...(1)$$

where $\xi(x, t)$ = particle displacement at a distance x from the disk at time t,

ρ_0 = density of undisturbed air,

c = velocity of sound,

$r = \frac{4}{3} \frac{\mu \omega^2}{c^2}$, μ = coefficient of viscosity, $\omega = 2\pi$ frequency.

Solve (1) subject to quiescence initially, and the boundary condition $\xi = \xi_0 \sin \omega t$ at $x = 0$, this being the displacement of the disk.

To deal with this problem rigorously, the third member of (1) should be replaced by $+\frac{4}{3}\mu \partial^3 \xi / \partial x^2 \partial t$, because the former pertains solely after the steady state has been attained. However, the analysis would then be very complicated, and the physical aspect would be masked. Moreover, to effect simplification, we shall use (1) and obtain an interesting approximation. The third term in (1) is analogous to the second term in the telegraph equation [12, (9), p. 237], $\mathbf{G} = 0$.

The analysis can be simplified by taking $\xi = \xi_0 e^{i\omega t}$ at $x = 0$ and finally selecting the imaginary part. Write $a = 1/c^2$, $b = r/\rho_0 c^2 = 4\mu\omega^2/3\rho_0 c^4$, then with

$$\xi(x, t) \fallingdotseq \phi(x, p),$$

the transform equation for quiescence initially is

$$\frac{d^2\phi}{dx^2} - (ap^2 + bp)\phi = 0. \quad\ldots\ldots\ldots\ldots\ldots\ldots(2)$$

Let $ap^2 + bp = a[(p + \beta)^2 - \beta^2] = \lambda^2$, where $\beta = b/2a$, and (2) becomes

$$\frac{d^2\phi}{dx^2} - \lambda^2\phi = 0, \quad\ldots\ldots\ldots\ldots\ldots\ldots(3)$$

so

$$\phi = Ae^{-\lambda x} + Be^{\lambda x}. \quad\ldots\ldots\ldots\ldots\ldots(4)$$

Unidirectional transmission entails absence of the B term, so with $\xi = \xi_0 e^{i\omega t}$ at $x = 0$, $\phi = \xi_0 p/(p - i\omega) = A$, which leads to

$$\xi(x, t) \fallingdotseq \phi = \frac{\xi_0 p}{p - i\omega} e^{-\frac{x}{c}\sqrt{(p+\beta)^2 - \beta^2}}. \quad\ldots\ldots\ldots\ldots(5)$$

4·17. Inverse of (5) § 4·16.   By § 2·13 with $a = -i\omega$

$$e^{-i\omega t}\xi(x, t) \fallingdotseq \xi_0 e^{-\frac{x}{c}\sqrt{(p+\beta+i\omega)^2 - \beta^2}}. \quad\ldots\ldots\ldots\ldots(1)$$

Inverting (1) from the list, we obtain

$$e^{-i\omega t}\xi(x, t) = \xi_0\left[e^{-\frac{x}{c}(\beta+i\omega)} H\left(t - \frac{x}{c}\right)\right.$$
$$\left. + \frac{\beta x}{c}\int_{x/c}^t e^{-(\beta+i\omega)t}\frac{I_1(\beta\sqrt{t^2 - x^2/c^2})}{\sqrt{t^2 - x^2/c^2}}\, dt\right]. \quad\ldots\ldots(2)$$

Omitting $H(t - x/c)$, the presence of which is tacitly understood, the particle displacement at any point x at time t is the imaginary part of

$$\xi(x,\, t) = \xi_0 \left[e^{i\omega(t-x/c)}\, e^{-\frac{\beta x}{c}} + \right.$$

$$\left. + e^{i\omega t}\frac{\beta x}{c}\int_{x/c}^{t} e^{-(\beta+i\omega)t}\frac{I_1(\beta\sqrt{t^2 - x^2/c^2})}{\sqrt{t^2 - x^2/c^2}}\, dt \right]. \quad \ldots\ldots(3)$$

When t is large, the value of the integral is sensibly that with t infinite. Using (2) § 1·31, and the list, and writing $(\beta + i\omega)$ for p, gives

$$\frac{\beta x}{c}\int_{x/c}^{\infty} e^{-(\beta+i\omega)t}\frac{I_1(\beta\sqrt{t^2 - x^2/c^2})}{\sqrt{t^2 - x^2/c^2}}\, dt$$

$$= e^{-\frac{x}{c}\sqrt{(\beta+i\omega)^2-\beta^2}} - e^{-\frac{x}{c}(\beta+i\omega)}, \quad \ldots\ldots\ldots\ldots\ldots(4)$$

$$= e^{-\frac{x}{c}(\beta+i\omega)\sqrt{1-(\beta/\beta+i\omega)^2}} - e^{-\frac{x}{c}(\beta+i\omega)}, \quad \ldots\ldots\ldots(5)$$

or $\qquad I \simeq e^{-\frac{x}{c}(\beta+i\omega)[1-\beta^2/2(\beta+i\omega)^2]} - e^{-\frac{x}{c}(\beta+i\omega)}, \quad \ldots\ldots\ldots(6)$

if $\omega \gg \beta$. Reducing still further and multiplying by $e^{i\omega t}$ yields

$$e^{i\omega t}I \simeq e^{-\frac{x\beta}{c}}\left[e^{i\omega\left\{t-\frac{x}{c}(1+\beta^2/2\omega^2)\right\}} - e^{i\omega(t-x/c)} \right]. \quad \ldots\ldots\ldots(7)$$

Inserting the value of $e^{i\omega t}I$ from (7) into (3) leads to

$$\xi(x,\, t) \simeq \xi_0 e^{-\frac{\beta x}{c}}\left[e^{i\omega\left\{t-\frac{x}{c}(1+\beta^2/2\omega^2)\right\}} \right], \quad \ldots\ldots\ldots(8)$$

when t is large enough. Taking the imaginary part, the particle displacement at $x,\, t$ [for $\xi = \xi_0 \sin \omega t$ at $x = 0$], approximates to

$$\xi(x,\, t) = \xi_0 e^{-\frac{\beta x}{c}}\sin\left[\omega\left\{t - \frac{x}{c}(1 + \beta^2/2\omega^2)\right\} \right]. \quad \ldots\ldots\ldots(9)$$

Compared with sound transmission in a loss-free medium,* the effect of viscosity is to

    (a) introduce the attenuation factor $e^{-\beta x/c}$;

    (b) reduce the *phase* velocity in the ratio
$$(1 + \beta^2/2\omega^2)^{-1} \simeq (1 - \beta^2/2\omega^2)/1.$$

* In this case $r = 0$ so $\beta = 0$ and $\xi(x,\, t) = \xi_0 \sin \omega(t - x/c)$, i.e. the attenuation is zero and the phase velocity equal to c the velocity of sound, independent of ω.

Thus the amplitude of the sound waves is attenuated exponentially with increasing distance x, while the phase velocity is reduced by an amount $\beta^2 c/2\omega^2 = 2\mu^2 k^2/9\rho_0^2 c$, where $k = \omega/c = 2\pi/\text{wave}$ length. The necessary check tests are left for the reader as exercises.

4·18. Compressional shock waves [ref. 2].

A very long uniform rigid channel of rectangular cross-section, as illustrated in Fig. 11, contains liquid to a depth h. At $t = 0$ a rigid vertical

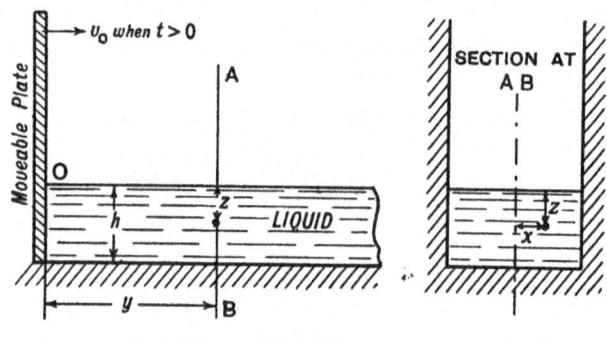

Fig. 11.

flat plate at the end $y = 0$ commences to move with constant horizontal velocity v_0. Neglecting losses, determine the velocity potential and excess pressure at any point of the liquid when $t > 0$. Assume the motion to be such that squares and higher powers of the velocity components can be disregarded. This problem is equivalent to that where water with a free surface impinges on a flat vertical wall.

Let $\phi =$ velocity potential at any point x, y, z in the liquid ; *

$\rho =$ density of liquid ;

$c =$ velocity of compressional waves ;

$p_1 = \rho\, \partial\phi/\partial t =$ excess pressure at x, y, z ;

$u_y =$ horizontal component of velocity at x, y, z ;

$u_z =$ vertical component of velocity at x, y, z.

* ϕ is the standard symbol for v.p. and will not be used for the L.T. in this example. p_1 is used for excess pressure to avoid confusion with the transform variable p.

Then by definition,

$$u_y = - \partial\phi/\partial y, \quad \text{................................(1)}$$

$$u_z = - \partial\phi/\partial z. \quad \text{................................(2)}$$

The v.p. must satisfy

1°. The wave equation, namely,

$$\frac{\partial^2\phi}{\partial x^2} + \frac{\partial^2\phi}{\partial y^2} + \frac{\partial^2\phi}{\partial z^2} = \frac{1}{c^2}\frac{\partial^2\phi}{\partial t^2}, \quad \text{....................(3)}$$

at all points (x, y, z) of the liquid ;

2°. the initial conditions, which correspond to quiescence ;

3°. the boundary conditions. These are

(a) $u_y = - \partial\phi/\partial y = v_0$, when $y = 0$ ⎫ constant velocity of end
(b) $u_z = - \partial\phi/\partial z = 0$, when $z = h$ ⎬ plate, bottom of channel
(c) $p_1 = \rho\,\partial\phi/\partial t = 0$, when $z = 0$ ⎭ stationary, excess pressure at free surface, zero.

For given values of y, z, the v.p. is constant for all x over the section, so $\partial\phi/\partial x = 0$. Hence the equation to be solved subject to the foregoing conditions is

$$\frac{\partial^2\phi}{\partial y^2} + \frac{\partial^2\phi}{\partial z^2} = \frac{1}{c^2}\frac{\partial^2\phi}{\partial t^2}. \quad \text{....................(4)}$$

4·19. Solution of (4) § 4·18. If we assume that

$$\phi(y, z, t) = \Sigma \sin\frac{\pi r z}{2h}\, f_r(y, t),$$

the third boundary condition is satisfied, as also is the second, provided r is an *odd* positive integer. Then with $f_r(y, t) \Rightarrow \psi_r(y, p)$, by (4) § 4·18 with $r_1 = \pi r/2h$, the transform equation is

$$\frac{d^2\psi_r}{dy^2} - (r_1^2 + p^2/c^2)\psi_r = 0, \quad \text{....................(1)}$$

provided that term by term differentiation of ϕ with respect to y, z is valid.

Assuming unidirectional propagation, the formal solution of (1) for ψ_r is

$$\psi_r = A_r e^{-\frac{y}{c}\sqrt{p^2 + c^2 r_1^2}}, \quad \text{....................(2)}$$

where A_r is a function of p. Thus

$$\phi_r = A_r \sin \frac{\pi r z}{2h} e^{-\frac{y}{c}\sqrt{p^2+c^2r_1^2}}, \qquad \dots\dots\dots\dots(3)$$

and

$$\phi = \Sigma A_r \sin \frac{\pi r z}{2h} e^{-\frac{y}{c}\sqrt{p^2+c^2r_1^2}}. \qquad \dots\dots\dots\dots(4)$$

To determine A_r, we have by boundary condition (a) $3°$ § 4·18,

$$-\left(\frac{\partial\phi}{\partial y}\right)_{y=0} = \frac{1}{c}\Sigma A_r \sqrt{p^2+c^2r_1^2}\,\sin\frac{\pi r z}{2h} = v_0, \qquad \dots\dots\dots(5)$$

provided differentiation under the integral sign in

$$\int_{y/c}^{\infty} e^{-pt}\phi(y,\, z,\, t)\,dt$$

is valid. Assuming this to hold, we obtain

$$\Sigma A_r \sqrt{p^2+c^2r_1^2}\,\sin\frac{\pi r z}{2h} = cv_0. \qquad \dots\dots\dots\dots(6)$$

Multiplying both sides of (6) by $\sin\dfrac{\pi n z}{2h}$, n an odd positive integer, and integrating with respect to z from 0 to h, all terms on the l.h.s. vanish except when $n = r$. Thus

$$A_r \int_0^h \sin^2\frac{\pi r z}{2h}\,dz = \frac{cv_0}{\sqrt{p^2+c^2r_1^2}}\int_0^h \sin\frac{\pi r z}{2h}\,dz, \qquad \dots\dots\dots(7)$$

so

$$A_r = 4cv_0/\pi r\sqrt{p^2+c^2r_1^2}. \qquad \dots\dots\dots\dots(8)$$

Substituting from (8) into (4), we obtain

$$\phi(y,\, z,\, t) = \frac{4cv_0}{\pi}\Sigma\frac{1}{r}\cdot\sin\frac{\pi r z}{2h}\frac{e^{-\frac{y}{c}\sqrt{p^2+c^2r_1^2}}}{\sqrt{p^2+c^2r_1^2}}, \qquad \dots\dots\dots(9)$$

so by aid of the list we find that the v.p. at any point y, z in a cross-section is

$$\phi = \frac{4cv_0}{\pi}\Sigma\frac{1}{r}\cdot\sin\frac{\pi r z}{2h}\int_{y/c}^t J_0(cr_1\sqrt{t^2 - y^2/c^2})\,dt. \qquad \dots\dots(10)$$

The pressure corresponding to (10) is

$$\rho \frac{\partial \phi}{\partial t} = \frac{4\rho c v_0}{\pi} \Sigma \frac{1}{r} \cdot \sin \frac{\pi r z}{2h} \cdot J_0(a\sqrt{t^2 - y^2/c^2}), \quad \ldots\ldots(11)$$

where $a = cr_1$. It is left to the reader to verify that

1. (10) satisfies (4) § 4·18, the initial, and the boundary conditions ;

2. Term by term differentiation of ϕ with respect to y, z, in (5) and in (10) is valid (see § 7, Appendix II) ;

3. Term by term correspondence in (9), (10) is valid ;

4. (10) may be differentiated term by term ;

5. $\dfrac{\partial^2}{\partial y^2} \displaystyle\int_{y/c}^{\infty} e^{-pt}\phi(y, z, t)dt = \int_{y/c}^{\infty} e^{-pt}\dfrac{\partial^2 \phi}{\partial y^2} dt$;

6. $\dfrac{\partial^2}{\partial z^2} \displaystyle\int_{y/c}^{\infty} e^{-pt}\phi(y, z, t)dt = \int_{y/c}^{\infty} e^{-pt}\dfrac{\partial^2 \phi}{\partial z^2} dt.$

For additional parts of the analysis and graphs of the wave form, reference should be made to [2].

4·191. The function $J_0(a\sqrt{t^2 - y^2/c^2})$. The time taken for a disturbance to travel from the origin to a point distant y therefrom, with velocity c, is $t_1 = y/c$. Thus in our problem the above function does not exist when $t < y/c$. Moreover, in a strict sense, it should be written $J_0(a\sqrt{t^2 - y^2/c^2})H(t - y/c)$ [see footnote to (1) § 1·31]. Now $J_0(a\sqrt{t^2 - y^2/c^2}) = J_0(at\sqrt{1 - y^2/c^2t^2})$, so if y is fixed, $\sqrt{1 - y^2/c^2t^2} < 1$ for $t > y/c$, and $\to 1$ as $t \to +\infty$. Thus for any value of y the graph of this function lags—so to speak—in point of time behind that of $J_0(at)$. The first zero of $J_0(at)$ occurs when $t \simeq 2\cdot405/a$, whilst that of $J_0(a\sqrt{t^2 - y^2/c^2})$ occurs when $a\sqrt{t^2 - y^2/c^2} \simeq 2\cdot405$ or $t \simeq \sqrt{\dfrac{2\cdot405^2}{a^2} + \dfrac{y^2}{c^2}} > 2\cdot405/a$. A similar argument pertains to all the zeros. When t is large enough compared with y/c, $\sqrt{t^2 - y^2/c^2} \sim t$, and the graphs of the two functions are almost coincident. The reader should plot $J_0(t)$ and $J_0(\sqrt{t^2 - y^2/c^2})$ for $y = 100$, 1000, 10,000, $c = 4000$, using the same coordinate axes for both graphs. Tabular values of $J_0(t)$ are given in [11].

4·21. Simultaneous partial linear D.E. with constant coefficients. Fig. 12a is a schematic representation of a uniform electrical transmission line or a loaded submarine telegraph cable having resistance **R**, inductance **L**, capacitance

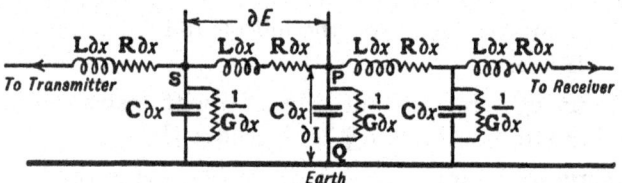

Fig. 12a.—Schematic diagram of electrical meshes equivalent to a uniform transmission line or a uniformly loaded submarine cable. Starting with a length δx, this is considered to $\rightarrow +0$ as the line $\rightarrow$ uniformity, i.e. becomes meshless.

$Z = Lp + R$, the transform of the series impedance of unit length ;

$Y = Cp + G$, the transform of the shunt admittance of unit length ;

$1/G\partial x$ is the resistance across the capacitance $C\partial x$; $G\partial x$ is its admittance ;

$\partial E = -\left(L\dfrac{\partial}{\partial t} + R\right) I_x \, \partial x$, the p.d. across SP, I_x being the current at P ;

$\partial I = -\left(C\dfrac{\partial}{\partial t} + G\right) E_x \, \partial x$, the current in PQ, E_x being the p.d. across PQ.

C, and leakance **G**, all assumed constant and reckoned per unit length (1 nautical mile of 6087 ft. in submarine cables). At any time $t > 0$, E and I represent the potential difference between a point distant x from a convenient origin (see Fig. 12b) and earth, and the current at x, respectively. Then at x the p.d. across the length ∂x between S and P is $-\partial E = [L\partial x](\partial I/\partial t) + [R\partial x] I$, so

$$-\frac{\partial E}{\partial x} = L\frac{\partial I}{\partial t} + RI = \left(L\frac{\partial}{\partial t} + R\right) I, \quad \ldots\ldots\ldots\ldots(1)$$

the negative sign indicating that E decreases as x increases. The current in PQ, by virtue of the capacitance and leakance of a length ∂x, is $-\partial I = [C\,\partial x](\partial E/\partial t) + [G\,\partial x] E$, so

$$-\frac{\partial I}{\partial x} = C\frac{\partial E}{\partial t} + GE = \left(C\frac{\partial}{\partial t} + G\right) E. \quad \ldots\ldots\ldots\ldots(2)$$

We consider first the so-called unloaded submarine telegraph cable, where **L** and **G** may be neglected. Then from (1), (2) we have

$$-\frac{\partial E}{\partial x} = \mathbf{R}I, \qquad\dotfill(3)$$

and

$$-\frac{\partial I}{\partial x} = \mathbf{C}\frac{\partial E}{\partial t}. \qquad\dotfill(4)$$

We shall take quiescent conditions initially, so that

$$E = 0, \; t < 0, \; 0 \leqslant x \;;$$
$$I = 0, \; t < 0, \; 0 \leqslant x \;;$$

and assume the cable to be very long. As in § 4·13, if the length is finite but great enough, the influence of the energy reflected from the far end upon E and I at an intermediate point x can be disregarded when x is sufficiently remote from that end, assumed to be open.* The transform equations corresponding to (3), (4) are by (8) § 2·14, (since the Σ term is zero in virtue of quiescence initially) ;

$$\frac{d\varphi}{dx} + \mathbf{R}\phi = 0, \qquad\dotfill(5)$$

and

$$\frac{d\phi}{dx} + \mathbf{C}p\varphi = 0, \qquad\dotfill(6)$$

where $E(x, t) \Rightarrow \varphi(x, p)$, $I(x, t) \Rightarrow \phi(x, p)$.† As in § 4·11 we assume E and I to be functions such that the order of differentiation and integration may be inverted, i.e.

$$\frac{d\varphi}{dx} = p\frac{\partial}{\partial x}\int_0^\infty e^{-pt}E \, dt = p\int_0^\infty e^{-pt}\left(\frac{\partial E}{\partial x}\right) dt, \qquad\dots\dots(7)$$

and

$$\frac{d\phi}{dx} = p\frac{\partial}{\partial x}\int_0^\infty e^{-pt}I \, dt = p\int_0^\infty e^{-pt}\left(\frac{\partial I}{dx}\right). dt. \qquad\dots\dots(8)$$

The validity of (7), (8) may be checked when E and I have been determined. Differentiating (5) with respect to x, and substituting the value of $d\phi/dx$ from (6) therein, we get

$$\frac{d^2\varphi}{dx^2} - \lambda^2\varphi = 0, \qquad\dotfill(9)$$

with $\lambda^2 = p\mathbf{C}\mathbf{R}$. If at time $t = 0$ a battery of voltage E_0 is applied between the cable at $x = 0$ and earth, as illustrated

---

* Insulated from earth.

† For brevity we shall frequently write E, I, or E_x, I_x to signify functions of (x, t).

schematically in Fig. 12b, electrical energy will be propagated along the cable towards the far end, there being (by hypothesis)

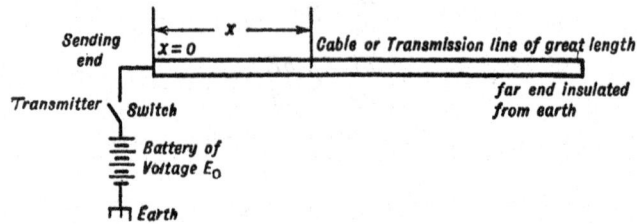

FIG. 12b. Schematic diagram of uniform electrical transmission line or submarine cable of great length, the far end being 'open', i.e. insulated from earth.

negligible reflection effect at x. The corresponding boundary conditions are

(a) $E\rightarrow0$, $x\rightarrow\infty$, $t>0$;

(b) $E=E_0$ (a constant), $x=0$, $t>0$, (see Fig. 13). This

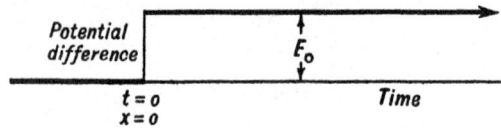

FIG. 13.—Diagram illustrating form of p.d. applied to sending end of submarine cable.

notation signifies that a p.d. E_0 is applied at $x=0$ when $t=0$, and maintained indefinitely.

The formal solution of (9) is

$$\varphi = Ae^{-\lambda x} + Be^{\lambda x}, \quad\quad\quad\quad\quad (10)$$

or
$$\varphi = A_1 \sinh \lambda x + B_1 \cosh \lambda x, \quad\quad\quad (11)$$

whichever is the better suited to the boundary conditions.

In virtue of the first boundary condition, it follows that $B=0$.

For the second boundary condition, since $\varphi = p \int_0^\infty e^{-pt} E\, dt$, where $E = E_0$ a constant, at $x=0$, we have $\varphi = E_0$. Thus $A = E_0$, and since $B=0$, (10) becomes

$$\varphi = E_0 e^{-\lambda x} = E_0 e^{-x\sqrt{p\mathbf{C}\mathbf{R}}}. \quad\quad\quad\quad (12)$$

By (5) and (12), we obtain

$$\phi = \lambda E_0 e^{-\lambda x}/\mathbf{R} = E_0 \sqrt{(p\mathbf{C}/\mathbf{R})}\, e^{-x\sqrt{p\mathbf{C}\mathbf{R}}}. \quad\quad\quad (13)$$

Formulae (12), (13) are, respectively, the *transform* solutions of equations (3), (4) for E_x and I_x. We have now to invert these formulae, i.e., obtain the corresponding functions of t.

4·211. Inversion of φ and ϕ in § 4·21. From the list, for (12) § 4·21 we have

$$\varphi = E_0 e^{-x\sqrt{p\mathbf{CR}}} \Subset E_0 \operatorname{erfc} \tfrac{1}{2}x\sqrt{(\mathbf{CR}/t)} = E_x; \quad \dots\dots\dots(1)$$

and for (13) § 4·21

$$\phi = E_0\sqrt{(p\mathbf{C}/\mathbf{R})}\,e^{-x\sqrt{p\mathbf{CR}}} \Subset E_0\sqrt{(\mathbf{C}/\pi t\mathbf{R})}\,e^{-x^2\mathbf{CR}/4t} = I_x. \quad\dots\dots(2)$$

Following the procedure given in §§ 4·111, 4·112, the reader will have no difficulty in showing that (1), (2) satisfy

1° Equations (3), (4) § 4·21,

2° The initial conditions,

3° The boundary conditions,

4° That $e^{-p_0 t}\begin{Bmatrix}E(x,\,t)\\ I(x,\,t)\end{Bmatrix}$ is bounded and continuous in $t>0$, $x>0$, $p_0>0$.

We now investigate the validity of the changed order of differentiation and integration in (7), (8) § 4·21.

4·212. Validity of (7), (8) § 4·21. By (1) § 4·211.

$$p\frac{\partial}{\partial x}\int_0^\infty e^{-pt}E\,dt = E_0\frac{\partial}{\partial x}e^{-x\sqrt{ap}} = -E_0\sqrt{ap}\,e^{-x\sqrt{ap}}, \quad\dots\dots(1)$$

where $a = \mathbf{CR}$. Also

$$p\int_0^\infty e^{-pt}\left(\frac{\partial E}{\partial x}\right)dt = -E_0\sqrt{a/\pi}\,p\int_0^\infty e^{-pt-ax^2/4t}t^{-1/2}\,dt, \quad\dots\dots(2)$$

$$= -E_0\sqrt{ap}\,e^{-x\sqrt{ap}}, \quad\dots\dots\dots\dots\dots(3)$$

by the list. In virtue of the identity of (1), (3), the changed order in (7) § 4·21 is valid. By (2) § 4·211

$$p\frac{\partial}{\partial x}\int_0^\infty e^{-pt}I\,dt = E_0\sqrt{p\mathbf{C}/\mathbf{R}}\,\frac{\partial}{\partial x}e^{-x\sqrt{ap}} = -E_0\mathbf{C}pe^{-x\sqrt{ap}}\dots(4)$$

Also

$$p\int_0^\infty e^{-pt}\left(\frac{\partial I}{\partial x}\right)dt = -\frac{E_0}{2}\sqrt{\mathbf{C}/\pi\mathbf{R}}\,axp\int_0^\infty e^{-pt-ax^2/4t}\,t^{-3/2}\,dt, \dots(5)$$

$$= -E_0\mathbf{C}pe^{-x\sqrt{ap}}, \quad\dots\dots\dots\dots\dots\dots(6)$$

provided $x>0$, so (8) § 4·21 is valid under this condition. As $x \to +0$ in (5), the integral does *not* converge uniformly, and when $x=0$ it diverges at the origin $t=0$. Hence although E and I in (1), (2) § 4·211 satisfy (4) § 4·21 for $x \geqslant 0$, the transform equation

$$-p \int_0^\infty e^{-pt} \left(\frac{\partial I}{\partial x} \right) dt = \mathbf{C}p \int_0^\infty e^{-pt} \left(\frac{\partial E}{\partial t} \right) dt, \quad \ldots \ldots \ldots (7)$$

is not true for $x=0$, since $[\partial I / \partial x]_{x=0}$ has no L.T. This case is akin to that in § 1·231, where $\phi(p)/p = p/(p^2+1)^{3/2}$ is continuous, but the integral diverges when $p=0$. In the above instance, $\phi(p)/p = ke^{-x\sqrt{ap}}$ is continuous at $x=0$, but integral (5) diverges.

4·213. The function (2) § 4·211. When $x=0$, this takes the form $t^{-1/2}$, so the current into the sending end of the cable is infinite in a mathematical sense. In practice the input current is limited by auxiliary apparatus associated with the transmitter. When $x>0$, I_x has a maximum value of

$$\sqrt{2/\pi}\,(E_0/\mathbf{R})e^{-1/2}x^{-1} \text{ at } t = \tfrac{1}{2}x^2\mathbf{CR}.$$

The current starts from zero at $t=0$ with zero rate of rise, owing to the cable being charged through its own resistance. $\displaystyle\int_0^t I_x\,dt$ gives the quantity of electricity supplied by the battery. Taking $\mathbf{C} = 3{\cdot}5 \times 10^{-7}$ farad, $\mathbf{R} = 2{\cdot}5$ ohms in (2) § 4·211, the reader should calculate I_x and t for $x=1$, and $x=1000$ nautical miles.

4·31. Solution of (1), (2) § 4·21 when G and L are not negligible. The equations (1), (2), § 4·21 symbolise the so-called uniformly *loaded* cable or transmission line where inductance plays the major role in reducing signal distortion. With an unloaded cable (**L** negligible), considerable signal distortion occurs at ordinary speeds of signalling, and correction devices are needed at the receiving end to obtain legible characters. In a loaded cable, the central copper conductor is ' loaded ' or wrapped with a spiral of thin tape or wire, having a high initial magnetic permeability,* of the order 4000. By this means the

---

* If B is flux density and H magnetising force, the initial permeability is $\mu = \lim\limits_{H \to +0} B/H$.

inductance per unit length may be increased some 40 to 60 times that of an unloaded cable,* and for a given character legibility at the receiver, the speed of signalling may be increased several fold.

To solve equations (1), (2) § 4·21, we proceed as shown in that section.

The respective transform equations are

$$\frac{d\varphi}{dx} + \mathbf{Z}\phi = 0, \quad\dots\dots\dots\dots\dots\text{(1)}$$

and

$$\frac{d\phi}{dx} + \mathbf{Y}\varphi = 0, \quad\dots\dots\dots\dots\dots\text{(2)}$$

where $\mathbf{Z} = (\mathbf{L}p + \mathbf{R})$, the L.T. of the series impedance per unit length of line,

$\mathbf{Y} = (\mathbf{C}p + \mathbf{G})$, the L.T. of the shunt admittance per unit length of line,

(see Fig. 12a). Using quiescent conditions initially, we obtain ultimately the formal solution given at (10) § 4·21, except that now $\lambda^2 = (\mathbf{L}p + \mathbf{R})(\mathbf{C}p + \mathbf{G})$. Thus if $v = 1/\sqrt{\mathbf{LC}}$, $\mathbf{R}/\mathbf{L} = 2a$, $\mathbf{G}/\mathbf{C} = 2b$, we get $\lambda = v^{-1}\sqrt{(p+2a)(p+2b)}$.

The same boundary conditions as before lead to the results

$$\varphi = E_0 e^{-\frac{x}{v}\sqrt{(p+2a)(p+2b)}}, \quad\dots\dots\dots\dots\dots\text{(3)}$$

and

$$\phi = E_0\sqrt{\mathbf{C}/\mathbf{L}} \frac{(p+2b)e^{-\frac{x}{v}\sqrt{(p+2a)(p+2b)}}}{\sqrt{(p+2a)(p+2b)}}. \quad\dots\dots\dots\text{(4)}$$

As will be evident in § 4·311, it is preferable to leave (4) as shown, and not cancel out $\sqrt{p+2b}$.

The exponential index in (3), (4). In (7), (8) § 4·21, t represents time, so p has dimension t^{-1} for the index to be dimensionless. The parameters a, b, have the same dimension as p. Thus $\sqrt{(p+2a)(p+2b)}$ has dimension t^{-1}. Since x represents length and the index is to be dimensionless, it follows that

---

* In § 4·21 we put $\mathbf{L} = 0$ as an analytical expedient, so that theoretically the velocity is infinite. The *actual* value of $\mathbf{L}$ gives $v = 1/\sqrt{\mathbf{LC}}$ of the order 3×10^4 nautical miles per second or 6×10^9 cm. p.s., i.e. one-fifth the velocity of light. In a long unloaded cable the received current would not alter appreciably if $\mathbf{L}$ were zero.

$v = 1/\sqrt{LC}$ has dimensions lt^{-1} and, therefore, represents a velocity. Typical values for a loaded cable are $L = 10^{-1}$ henry, $C = 4 \times 10^{-7}$ farad per nautical mile. The corresponding velocity of propagation is approximately 5000 nautical miles per second or about 10^9 cm. per sec. This is roughly one-sixth the *actual*[*] velocity in an unloaded cable, and one-thirtieth that of light.

The finite velocity of propagation means that the current reaches a point distant x from the sending end at a time $t_1 = x/v$ *after* the transmitting key at that end is closed. Hence in the integrals corresponding to (7), (8) § 4·21 the lower limit must be x/v and not zero, but the boundary conditions are unaltered. Moreover, the functions representing E and I will take a form where the origin is moved to the point $t_1 = x/v$; and so far as our problem is concerned, they are both zero prior to that instant.

4·311. Inversion of (4) § 4·31 in terms of t.

If we put $(a + b) = \alpha$ and $(a - b) = \beta$, then

$$\lambda = v^{-1}\sqrt{(p + 2a)(p + 2b)} = v^{-1}\sqrt{(p + \alpha)^2 - \beta^2}, \quad \ldots\ldots\ldots(1)$$

so (4) § 4·31 becomes

$$\phi = E_0\sqrt{C/L} \, \frac{(p + 2b)e^{-(x/v)\sqrt{(p+\alpha)^2 - \beta^2}}}{\sqrt{(p + \alpha)^2 - \beta^2}} . \quad \ldots\ldots\ldots\ldots(2)$$

Consider the first term on the r.h.s. of (2), and write

$$e^{-\alpha t}f(x, t) = \frac{pe^{-(x/v)\sqrt{(p+\alpha)^2 - \beta^2}}}{\sqrt{(p + \alpha)^2 - \beta^2}} . \quad \ldots\ldots\ldots\ldots(3)$$

Then by § 2·13,

$$f(x, t) = \frac{pe^{-(x/v)\sqrt{p^2 - \beta^2}}}{\sqrt{p^2 - \beta^2}} \Subset I_0(\beta\sqrt{t^2 - x^2/v^2}), \quad \ldots\ldots\ldots(4)$$

$t > x/v$, from the list. Hence by (2)–(4) the inverse of the first term in (2) is

$$I_1(x, t) = E_0\sqrt{C/L} \, e^{-\alpha t}I_0(\beta\sqrt{t^2 - x^2/v^2}). \quad \ldots\ldots\ldots(5)$$

By § 2·151 the inverse of the second term in (2) is $2b$ times the integral of (5) taken between the limits x/v and t. Hence

[*] i.e. when $L \neq 0$.

the current at any point distant x from the sending end of the line is, for $t \geqslant x/v$,

$$I(x, t) = E_0\sqrt{C/L}\left[e^{-\alpha t}I_0(\beta\sqrt{t^2 - x^2/v^2})\right.$$

$$\left. + 2b\int_{x/v}^{t} e^{-\alpha t}I_0(\beta\sqrt{t^2 - x^2/v^2})dt\right]. \quad \ldots\ldots(6)$$

At the sending end $x = 0$, and when $t = 0$ the current rises precipitately to the value $E_0\sqrt{(C/L)}$, as illustrated in Fig. 14.

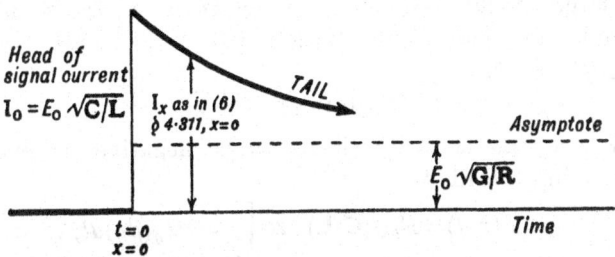

FIG. 14.—Current at sending end of uniform transmission line or loaded cable when $C/L > G/R$, i.e. $R/L > G/C$.

Comparing this with the result in § 4·211 we see that the effect of inductance is to limit the initial input current to the line. The quantity $\sqrt{(L/C)}$ is termed the input impedance of the line, and dimensionally it represents a resistance. Using the numerical data in § 4·31, its value is 500 ohms.

Formula (6) is valid for $t = x/v$, when we get

$$I_x = E_0\sqrt{(C/L)}e^{-\alpha x/v}, \quad \ldots\ldots\ldots\ldots\ldots\ldots(7)$$

since $I_0(0) = 1$. The physical interpretation of (7) is that the current travels down the line at velocity $v = 1/\sqrt{(LC)}$, with a vertical wave front, which is attenuated exponentially *en route*. The wave front is vertical since by hypothesis **L**, **R**, **C**, and **G** are *constant*. The signal which travels down the line may be considered at $x = 0$ to have a frequency spectrum extending from zero to infinity. The higher frequencies are essential to ensure a steep wave front. In practice **L** decreases while **R** increases with rise in frequency, and the wave front, say 3000 nautical miles from the sending end, is far from being vertical, although it is much steeper than that in an unloaded cable with

$L \simeq 0$. The front (signal head) is vertical at $x = 0$, but steadily tilts backwards * due to reduction in the higher frequencies with increase in x, as shown by the broken curve in Fig. 16.

In Fig. 14 the current *into* the line at $x = 0$ is shown to *decrease* asymptotically to the value $E_0 \sqrt{(\mathbf{G}/\mathbf{R})}$. This occurs when $\mathbf{R}/\mathbf{L} > \mathbf{G}/\mathbf{C}$, as is the case in practice. But if $\mathbf{G}/\mathbf{C} > \mathbf{R}/\mathbf{L}$ the current *increases* asymptotically to the above value. The final current † is due entirely to leakage through the insulation, and its value may be found from (6) in the following way. When t is large enough and much greater than x/v, the asymptotic formula for the Bessel function can be used, so with $u = \sqrt{(t^2 - x^2/v^2)}$,

$$I_0(\beta u) \sim e^{\beta u}/\sqrt{2\pi\beta u}. \qquad (8)$$

Now $-\alpha t + \beta u$ is easily shown to be negative, so making t infinite in (6) leads to

$$I(x, t) = E_0 \sqrt{(\mathbf{C}/\mathbf{L})} \cdot 2b \int_{x/v}^{\infty} e^{-\alpha t} I_0(\beta u) dt, \qquad (9)$$

$$= E_0 \sqrt{(\mathbf{C}/\mathbf{L})} \cdot 2b \, e^{-(x/v)\sqrt{\alpha^2 - \beta^2}} / \sqrt{\alpha^2 - \beta^2}$$

by the list
$$= E_0 \sqrt{(\mathbf{C}/\mathbf{L})} \sqrt{(b/a)} e^{-2(x/v)\sqrt{ab}}, \qquad (10)$$

$$= E_0 \sqrt{(\mathbf{G}/\mathbf{R})} e^{-x\sqrt{\mathbf{GR}}}. \qquad (11)$$

When $x = 0$ and $t \to +\infty$, the total leakage current of the line is

$$I(0, \infty) \to E_0 \sqrt{(\mathbf{G}/\mathbf{R})}. \qquad (12)$$

By using formula (11) § 13·41 in reference [12], it can be shown that for an open-ended (insulated) uniform line of length l,

$$I(0, \infty) = E_0 \sqrt{\mathbf{G}/\mathbf{R}} \tanh l\sqrt{\mathbf{GR}}. \qquad (13)$$

When $l\sqrt{\mathbf{GR}}$ is large enough,

$$\tanh l\sqrt{\mathbf{GR}} \sim 1, \qquad (14)$$

and (12) above is reproduced.

4·312. The function $y = I_0(\beta\sqrt{t^2 - x^2/v^2})$.

From a purely mathematical viewpoint, this function exists for all values of $\beta\sqrt{t^2 - x^2/v^2}$, whether real or complex. When β, t, x/v are real, the range of $\beta\sqrt{t^2 - x^2/v^2}$ is from $\pm i\infty$ to 0, and from 0 to $+\infty$,

* On a time basis.

† In a long line open (insulated from earth) at the far end. See re-. marks below (4) § 4·21.

according as $t <$ or $> x/v$. In our problem the variable is real, and the function commences at $t = x/v$. It is finitely discontinuous there and $I_0(0) = 1$. Consequently it ought to be expressed in the form

$$y = I_0(\beta\sqrt{t^2 - x^2/v^2})\,H(t - x/v), \quad\dots\dots\dots\dots(1)$$

where $H(t - x/v)$ is Heaviside's unit function with its origin at the point $t = x/v$. Then in (1) $y = 0$ for $t < x/v$, is discontinuous at $t = x/v$, and $y = I_0(\beta\sqrt{t^2 - x^2/v^2})$ when $t > x/v$. This applies also to the first term on the r.h.s. of (4) § 4·314, so it should read $e^{-\alpha x/v}H(t - x/v)$. The relationship between y and t is portrayed in Fig. 15, the value of t_1 being arbitrary. Using the numerical

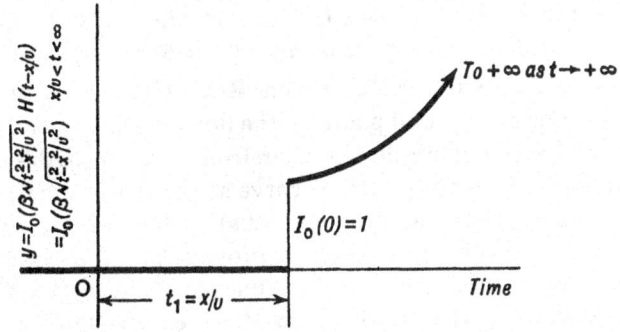

FIG. 15.—Graph of the Bessel function

$$y = I_0(\beta\sqrt{t^2 - x^2/v^2}) \left.\begin{array}{l} \\ \\ \end{array}\right\} \begin{array}{l} t > x/v \\ t < x/v. \end{array}$$
$$= 0$$

It is finitely discontinuous at $t_1 = x/v$, but positive, monotonic increasing and continuous in $t > x/v$, unbounded at $+\infty$.

data in §§ 4·31, 4·313 and tabular values of $I_0(x)$, the reader should plot (1) for various values of x, e.g. 0, 1000, 2000, 3000 nautical miles.

Differentiation of $y = I_0(\beta u)$, where $u = \sqrt{t^2 - x^2/v^2}$. Since y is discontinuous at $t = x/v$, it is indeterminate and *cannot be differentiated there.* But the function is continuous as $t \to +x/v$, so its first derivative may be found as a limiting case. (See Fig. 15 where the derivative is represented by the slope of the curve). Then

$$\frac{dy}{dt} = \frac{dI_0(\beta u)}{d(\beta u)} \cdot \frac{d(\beta u)}{dt} = \frac{\beta t}{u}\,I_0{}'(\beta u) = \beta^2 t\,\frac{I_1(\beta u)}{\beta u} \cdot \quad\dots\dots\dots(2)$$

Now as $t \to +x/v$, βu and $I_1(\beta u) \to +0$, but $I_1(\beta u)/\beta u \to 1/2$. [ref. 11]. Hence

$$\lim_{t \to +x/v} \frac{dy}{dt} = \beta^2 x/2v. \quad \ldots\ldots\ldots\ldots\ldots\ldots(3)$$

4·313. Discussion of (6) § 4·311. We have already considered the relationship between I_x and t at the origin $x = 0$ (see Fig. 14). As x increases, the head of the signal is attenuated exponentially, by virtue of the factor $e^{-\alpha x/v}$, and in a given transmission line, the shape of the ensuing part (or tail of the signal) depends upon x. The size of the head shrinks steadily compared with that of the tail, until the point $x = 2v/\beta$ is reached ($\beta > 0$). Thereafter the current-time relationship takes the form of the solid curve in Fig. 16, and the size of the tail exceeds that of the signal head. The current increases asymptotically to the value $E_0\sqrt{\mathbf{G}/\mathbf{R}}e^{-x\sqrt{\mathbf{GR}}}$. When $\mathbf{R/L} = \mathbf{G/C}$, $\beta = 0$ and the current-time curve at all points of the line (which are far enough from the open end for reflection therefrom to be negligible),* has the same shape as the p.d.-time curve at the sending end, i.e. at any point x it has the form $H(t - x/v)$. Moreover, if $1/\mathbf{LC}$ is constant, so also is the velocity of propagation, but the signals are now distortionless due to the influence of the large leakance $\mathbf{G}$ in preventing the size of the tail from exceeding that of the head.† The attenuation of the signals is so great, however, that the distortionless condition cannot be used in practice with the materials of construction now available. For the cable cited in § 4·31, $\mathbf{G}$ is 10^{-8} mho, $\mathbf{R}$ is 1·8 ohm per nautical mile, so $\mathbf{R/L} = 18$, whilst $\mathbf{G/C} = 1/40$. Thus in the attenuation index, namely, $\alpha = \frac{1}{2}[(\mathbf{R/L}) + (\mathbf{G/C})]$, $\mathbf{G/C}$ can be neglected in comparison with $\mathbf{R/L}$. But if $\mathbf{G}$ were increased 720-fold to obtain the distortionless criterion, α would be almost doubled. If $e^{-\alpha x/v}$ were 2×10^{-3} at a certain point of the line, with $\mathbf{G} = 10^{-8}$ mho, it would be 4×10^{-6} with a 720-fold value of $\mathbf{G}$. Thus the distortionless condition would entail a 500-fold reduction in signal head, which is impracticable.

* See footnote on p. 67.

† Current leaks from the central conductor through the insulation to earth. The leakage current decreases exponentially along the line: see (11) § 4·311. $\mathbf{G}$ may be regarded as the insulation conductance.

We shall now consider the preceding statements from an analytical standpoint. To obtain a rising current (solid curve of Fig. 16), the slope of the curve as $t \rightarrow +x/v$ must be > 0. Differentiating (6) § 4·311, we get

$$\frac{dI_x}{dt} = Ke^{-\alpha t}\left[\beta t\frac{I_1(\beta u)}{u} - (\alpha - 2b)I_0(\beta u)\right], \quad \ldots\ldots\ldots\ldots(1)$$

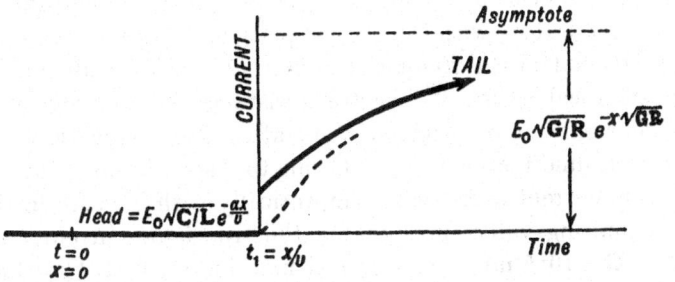

Fig. 16.—Form of current in uniform transmission line if $x > 2v/\beta$, when the p.d. at the sending end is $E_0H(t)$ as in Fig. 13.

where $\quad K = E_0\sqrt{C/L}, \quad u = \sqrt{t^2 - x^2/v^2}, \quad (\alpha - 2b) = \beta.$

Then $\qquad \lim_{t \rightarrow +\frac{x}{v}} \left(\frac{dI_x}{dt}\right) = Ke^{-\alpha x/v}\beta\left(\frac{\beta x}{2v} - 1\right), \quad \ldots\ldots\ldots\ldots(2)$

and for I_x to increase with increase in t, (2) must be > 0. Hence

$$x > 2v/\beta, \quad \ldots\ldots\ldots\ldots\ldots\ldots\ldots\ldots\ldots\ldots(3)$$

with $\beta > 0$, i.e. $\mathbf{R/L} > \mathbf{G/C}$. If $\mathbf{G} = 0$, (3) gives

$$x > \frac{4\sqrt{(\mathbf{L/C})}}{\mathbf{R}}, \quad \ldots\ldots\ldots\ldots\ldots\ldots(4)$$

which means that x must exceed four times the ratio of the input impedance to the resistance per unit length of the line. In the case treated above, $x > 1111$ nautical miles. When $\beta x/2v < 1$, (2) is negative, the current decreases with increase in time for a fixed x, so the signal head is larger than the tail.

If in (1) we take $\mathbf{R/L} = \mathbf{G/C}$, i.e. $\beta = 0$, dI_x/dt is zero for all values of $t > x/v$ and, therefore, the signal * is transmitted down the line without distortion of form, i.e. there is no change in type. Alternatively writing $\beta = 0$ in (6) § 4·311, $2b = \alpha$ and

* Of form $H(t)$ at the transmitter.

$I_x = E_0 \sqrt{(C/L)} e^{-\alpha x/v}$. Thus the current at a given x is constant for $t > x/v$, and has the form $H(t - x/v)$. The general case where the p.d. at the transmitter has the form $\xi(t)$, is treated in § 4·332.

Remarks on **G**. In practice **G** has a dual interpretation. (1) The effective leakance with sinusoidal current. This arises through " dielectric hysteresis " (the current being not quite in phase quadrature with the applied p.d.), and it is approximately proportional to the frequency $\omega/2\pi$. (2) The " conductance " or reciprocal of the " dielectric resistance " (the latter being p.d./current) at a specified time after the application of a constant unidirectional p.d. Owing to "absorption", the conduction current decreases asymptotically with increase in time to a limiting value, presumably the true ionic current. The value **G** $= 10^{-6}$ mho per nautical mile in [11, p. 131] is for a guttapercha dielectric at a frequency of 33 c.p.s. A value of **G** of this order would probably apply to the signal head (wave front) in the transient case, but for the " tail " of the transient, a value corresponding to the dielectric resistance is required, e.g. **G** $= 10^{-8}$ mho per nautical mile for guttapercha. For a synthetic dielectric, e.g. polythene, **G** would be much smaller than 10^{-8}. In modern cable telegraphy it would not be advantageous to increase **G**, since distortion is easily overcome by aid of correcting networks associated with the thermionic valve amplifiers at the receiving end of the cable.

4·314. Interpretation of (3) § 4·31 in terms of t. From (2) § 1·31, (3), and (4) § 4·311, we have

$$\int_{x/v}^{\infty} e^{-pt} \left[e^{-\alpha t} I_0 \{ \beta \sqrt{(t^2 - x^2/v^2)} \} \right] dt$$
$$= e^{-(x/v)\sqrt{\{(p+\alpha)^2 - \beta^2\}}} / \sqrt{\{(p+\alpha)^2 - \beta^2\}}. \quad \ldots\ldots(1)$$

Since the resulting integral converges uniformly (see §§ 12, 13, Appendix III for other conditions necessary) we may differentiate both sides of (1) with respect to x. Thus we obtain

$$e^{-(p+\alpha)(x/v)} + \frac{\beta x}{v} \int_{x/v}^{\infty} e^{-pt} \left[e^{-\alpha t} \frac{I_1 \{ \beta \sqrt{(t^2 - x^2/v^2)} \}}{\sqrt{(t^2 - x^2/v^2)}} \right] dt$$
$$= e^{-(x/v)\sqrt{\{(p+\alpha)^2 - \beta^2\}}}. \quad \ldots\ldots(2)$$

Inverting the l.h.s. of (2) using §§ 2·122, and 2·151, leads to the result

$$e^{-\alpha x/v} + \frac{\beta x}{v} \int_{x/v}^{t} e^{-\alpha t} \frac{I_1(\beta u)}{u} dt \rightleftharpoons e^{-\frac{x}{v}\sqrt{(p+\alpha)^2-\beta^2}}, \qquad \dots\dots\dots(3)$$

where it is to be understood that strictly the first member should be written $e^{-\alpha x/v}H(t-x/v)$. Hence by (3) § 4·31 and (3) above, the p.d. at any point x when $t>x/v$, is

$$E_x = E_0 \left[e^{-\alpha x/v} + \frac{\beta x}{v} \int_{x/v}^{t} e^{-\alpha t} \frac{I_1(\beta u)}{u} dt \right]. \qquad \dots\dots\dots(4)$$

This formula is illustrated graphically in Fig. 17.

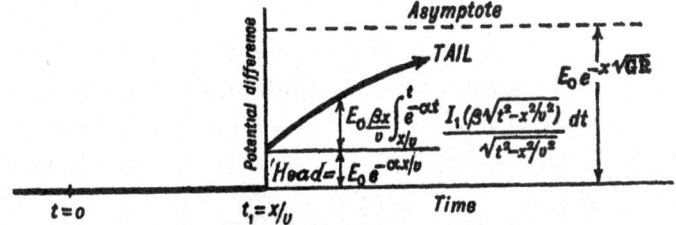

FIG. 17.—Illustrating relationship between p.d. and time in uniform transmission line or loaded submarine cable at a point distant x from the transmitter.

4·32. Proof that (6) § 4·311 and (4) § 4·314 are the appropriate solutions of (1), (2) § 4·21.

In this case proof that the differential equations are satisfied entails some heavy algebra involving Bessel functions, and need not be given here. The same remark applies to verification of the validity of the inverted order of integration in the integrals corresponding to (7), (8) § 4·21.* For the initial conditions, owing to the ordinary discontinuities in E, I, we let $t \to -0$. Then since (4) § 4·314 and (6) § 4·311 are zero prior to $t=0$, $E=I=0$ for all $x \geqslant 0$. Also if $t \to (x/v) - 0$, $E_x = I_x = 0$ at any x. For the boundary conditions, from (4) § 4·314 (a) $E \to 0$, $x \to \infty$, $0 < t < \infty$, (b) $E = E_0$, $x = 0$, $t > 0$. Thus both the initial and the boundary conditions are satisfied. Also E_x and I_x are bounded and continuous in $t > x/v$, $x \geqslant 0$; so also are $e^{-p_0 t}E_x$, $e^{-p_0 t}I_x$ for p_0 real > 0.

* In the present case the lower limit in each integral is x/v.

*4·321. Loaded cable with series capacitor C_S at sending end. In Fig. 12 b imagine C_s to be connected between the battery and the cable. Then, for quiescent initial conditions, the p.d. applied between $x=0$ and earth is E_0 less the p.d. across C_s. If $I(0, t) \Rightarrow \phi_0(p)$ is the current into the cable at $x=0$, the p.d. across C_s is

$$\frac{1}{C_s}\int_0^t I(0, t)dt \Rightarrow \phi_0/pC_s. \quad\quad\quad\dotfill(1)$$

Thus the p.d. applied between the cable and earth is

$$E(0, t) \Rightarrow (E_0 - \phi_0/pC_s), \quad\quad\quad\dotfill(2)$$

this being one boundary condition. The second b.c. is that

$$E(x, t) \rightarrow 0, \; x \rightarrow \infty, \; t > 0. \quad\quad\dotfill(3)$$

The L.T. of the current at any point distant x from the sending end of the cable is given by (13) § 4·21, provided we write $(\mathbf{L}p + \mathbf{R})$ for $\mathbf{R}$, $\lambda = \sqrt{\mathbf{LC}}\sqrt{(p+\alpha)^2 - \beta^2}$, and $(E_0 - \phi_0/pC_s)$ for E_0. Then we get

$$\phi = (E_0 - \phi_0/pC_s)\, \lambda e^{-\lambda x}/(\mathbf{L}p + \mathbf{R}). \quad\dotfill(4)$$

Putting $\lambda/(\mathbf{L}p + \mathbf{R}) = 1/Z_c$, where Z_c is the L.T. of the characteristic or surge impedance of the cable, and letting $x=0$, (4) becomes

$$\phi_0 = (E_0 - \phi_0/pC_s)/Z_c, \quad\quad\dotfill(5)$$

so
$$\phi_0 = E_0/(Z_c + 1/pC_s). \quad\quad\dotfill(6)$$

Substituting for ϕ_0 from (6) into (4) leads to the L.T. of the expression for the current, namely,

$$\phi = E_0 e^{-\lambda x}/(Z_c + 1/pC_s). \quad\quad\dotfill(7)$$

Since $\lambda = \sqrt{(\mathbf{L}p + \mathbf{R})(\mathbf{C}p + \mathbf{G})}$, we obtain

$$Z_c = \sqrt{\mathbf{L/C}}\sqrt{(p + 2a)(p + 2b)}$$

in the notation of §§ 4·31, 4·311. Inserting this in (7), and multiplying the numerator and denominator by $\sqrt{\mathbf{C/L}}(p + 2b)$, yields

$$\phi = \frac{E_0\sqrt{\mathbf{C/L}}(p + 2b)\,e^{-\lambda x}}{\sqrt{(p + 2a)(p + 2b)} + \rho(p + 2b)/p}, \quad\dotfill(8)$$

where $\rho = 1/C_s\sqrt{\mathbf{L/C}}$. Since $\sqrt{\mathbf{L/C}}$ has the same dimensions as a resistance, ρ has the dimension t^{-1} like $1/CR$ in $e^{-t/CR}$. Using (1) § 4·311 in (8) gives

$$I(x, t) \Rightarrow \phi = E_0\sqrt{\mathbf{C/L}} \; \frac{(p+2b)e^{-\frac{x}{v}\sqrt{(p+a)^2-\beta^2}}}{\sqrt{(p+\alpha)^2-\beta^2}+\rho(p+2b)/p} \cdot \quad\ldots\ldots(9)$$

To simplify the analysis which follows, we shall assume that **G** is negligible (see § 4·313). Then $b=0$, and $a=\alpha=\beta$, so (9) may be written

$$\frac{I(x, t)}{E_0\sqrt{\mathbf{C/L}}} \Rightarrow \frac{pe^{-\frac{x}{v}\sqrt{(p+a)^2-a^2}}}{\sqrt{(p+a)^2-a^2}+\rho} \cdot \quad\ldots\ldots\ldots\ldots(10)$$

Applying the theorem $e^{at}f(t) \Rightarrow \left(\dfrac{p}{p-a}\right)\phi(p-a)$ to (10) leads to

$$f(x, t) = \frac{e^{at}I(x, t)}{E_0\sqrt{\mathbf{C/L}}} \Rightarrow \frac{pe^{-\frac{x}{v}s}}{(s+\rho)}, \quad\ldots\ldots\ldots\ldots(11)$$

with $s=\sqrt{p^2-a^2}$.

*4·322. Inversion of (11) § 4·321. We may write

$$f(x, t) \Rightarrow \frac{pe^{-\frac{x}{v}s}}{s} - \frac{\rho p e^{-\frac{x}{v}s}}{s(s+\rho)} \cdot \quad\ldots\ldots\ldots\ldots(1)$$

From the list

$$I_0(a\sqrt{t^2-x^2/v^2}) \Rightarrow pe^{-\frac{x}{v}s}/s, \quad\ldots\ldots\ldots\ldots(2)$$

so we have now to invert the third member of (1). Using w for t *pro tem.*, the list gives $e^{-\rho w} \Rightarrow p/(p+\rho)$, and on applying the shift theorem we obtain

$$\rho e^{-\rho(w-x/v)} \Rightarrow \rho p e^{-\frac{x}{v}p}/(p+\rho). \quad\ldots\ldots\ldots\ldots(3)$$

If to (3) we apply (3) § 2·254, the third member of (1) is reproduced. Consequently in (3) § 2·254

$$f(w) = \rho e^{-\rho(w-x/v)}, \quad\ldots\ldots\ldots\ldots(4)$$

so it follows that

$$\rho e^{\rho x/v}\int_{x/v}^{t} e^{-\rho w}I_0(a\sqrt{t^2-w^2})\,dw \Rightarrow \rho p e^{-\frac{x}{v}s}/s(s+\rho), \quad\ldots\ldots(5)$$

the lower limit being x/v, since the current is zero for $t < x/v$. Hence from (11) § 4·321, (1), (2), and (5) above, we obtain

$$I(x,\,t) = E_0\sqrt{C/L}\,e^{-at}\Big[I_0(a\sqrt{t^2 - x^2/v^2})$$

$$-\,\rho e^{\rho x/v}\int_{x/v}^{t} e^{-\rho w}I_0(a\sqrt{t^2 - w^2})\,dw\Big]. \quad\ldots\ldots\ldots(6)$$

To find $I(x,\,t)$ when C_s is short-circuited (before the battery is connected), we put $C_s = \infty$. Then $\rho = 0$, and from (6) we get

$$I(x,\,t) = E_0\sqrt{C/L}\,e^{-at}I_0(a\sqrt{t^2 - x^2/v^2}). \quad\ldots\ldots\ldots\ldots(7)$$

Moreover, the second member on the r.h.s. of (6) represents the reduction in current caused by C_s. The requisite checking akin to that in § 4·32 is left to the reader.

*4·323. The solution when G $\neq$ 0. Using the numerical data in § 4·313 we get $a = 9$, $b = 0\cdot0125$, so $\alpha = 9\cdot0125$ and $\beta = 8\cdot9875$. Thus $\alpha \simeq \beta$, and this enables an adequate approximate solution to be found, the accuracy of which increases the smaller is $t > x/v$. Applying the theorem above (11) § 4·321 to (9) § 4·321, with α for a, and using $(2b - \alpha) = -\beta$, $s_1 = \sqrt{p^2 - \beta^2}$, we obtain

$$\frac{e^{\alpha t}I(x,\,t)}{E_0\sqrt{C/L}} = \Big(\frac{p - \beta}{p - \alpha}\Big)pe^{-\frac{x}{v}s_1}\Big/[s_1 + \rho(p - \beta)/(p - \alpha)] \quad\ldots\ldots(1)$$

$$= \frac{pe^{-\frac{x}{v}s_1}}{[s_1 + \rho(p - \beta)/(p - \alpha)]}\{1 + 2b/(p - \alpha)\}. \quad\ldots\ldots(2)$$

Taking $\alpha \simeq \beta$, (2) may be written

$$f(x,\,t) = \frac{pe^{-\frac{x}{v}s_1}}{(s_1 + \rho)}\{1 + 2b/(p - \alpha)\}. \quad\ldots\ldots\ldots\ldots\ldots\ldots(3)$$

Apart from the external multiplier, the inverse of the first term on the r.h.s. of (3) is given by (6) § 4·322, if α (or β) is written for a. The inverse of the second term in (3) may be obtained from that of the first by applying the product theorem. It is left for the reader to derive the result

$$I(x,\,t) = E_0\sqrt{C/L}\,\Big\{e^{-\alpha t}I_0(\beta\sqrt{t^2 - x^2/v^2})$$

$$-\,\rho e^{-\alpha t + \rho x/v}\int_{x/v}^{t} e^{-\rho w}I_0(\beta\sqrt{t^2 - w^2})\,dw +$$

$$+ 2b \int_{x/v}^{t} e^{-\alpha\mu} \Big[I_0 (\beta \sqrt{\mu^2 - x^2/v^2})$$

$$- \rho e^{\rho x/v} \int_{x/v}^{\mu} e^{-\rho w} I_0 (\beta \sqrt{\mu^2 - w^2})\, dw \Big]\, d\mu \Big\} . \quad \ldots \ldots (4)$$

When $C_s = \infty$ (the analy :al equivalent of a short-circuited capacitor), $\rho = 0$, and (4) reduces to (6) § 4·311, except that now $\alpha \simeq \beta$. Moreover the sum of the two members of (4), having ρ as a factor, represents the reduction in current caused by the capacitor.

*4·33. Boundary conditions as arbitrary functions of t
Referring to (10) § 4·21, let the boundary conditions for a very long line be (a) $E \to 0$, $x \to \infty$, $t > 0$; (b) $E = \xi_2(t)$, $x = 0$, $t > 0$. That B is zero follows immediately from (a). For (b) we have

$$\phi_2 = p \int_0^\infty e^{-pt} \xi_2(t)\, dt, \quad \ldots \ldots \ldots \ldots \ldots (1)$$

or $$\xi_2(t) \rightleftharpoons \phi_2(p). \quad \ldots \ldots \ldots \ldots \ldots (2)$$

Using this in (10) § 4·21, with $B = 0$, yields

$$\varphi = \phi_2 e^{-\lambda x}, \quad \ldots \ldots \ldots \ldots \ldots (3)$$

so that in (12) § 4·21 E_0 is replaced by the L.T. of $\xi_2(t)$, the applied p.d.

*4·331. Suppose that the boundary conditions for a limited length of line l are as follows :

(a) $E_1 = \xi_1(t) \rightleftharpoons \phi_1(p)$, $x = l$, $t > 0$;

(b) $E_2 = \xi_2(t) \rightleftharpoons \phi_2(p)$, $x = 0$, $t > 0$.

Then substituting (a) and (b) in (11) § 4·21 leads to

$$A_1 \sinh \lambda l + B_1 \cosh \lambda l = \phi_1, \quad \ldots \ldots \ldots \ldots (1)$$

$$B_1 = \phi_2. \quad \ldots \ldots \ldots \ldots (2)$$

Solving (1), (2) the arbitrary constants are obtained as functions of p, namely,

$$A_1 = (\phi_1 - \phi_2 \cosh \lambda l)/\sinh \lambda l, \quad \ldots \ldots \ldots \ldots (3)$$

$$B_1 = \phi_2. \quad \ldots \ldots \ldots \ldots (4)$$

The results in § 4·33 and the present section hold for the general value of λ, namely, $\sqrt{\{(\mathbf{L}p + \mathbf{R})(\mathbf{C}p + \mathbf{G})\}}$.

*4·332. Example. A potential difference represented by $\xi_2(t)$ is applied at $x=0$, $t=0$ to a very long loaded cable initially quiescent. Determine the p.d. and current at any point distant x from the transmitter. [See Fig. 12b where the p.d. due to the battery is now replaced by $\xi_2(t)$.]

1°. *The p.d.* By (3) § 4·33, with $\xi_2(t) \Rightarrow \phi_2(p)$,

$$\varphi = \phi_2 e^{-\lambda x}, \quad \dots\dots\dots\dots\dots(1)$$

λ being given in § 4·331. We apply the product theorem § 2·241 with

$$\phi_1 = \frac{pe^{-\lambda x}}{\lambda} \Subset ve^{-\alpha t}I_0(\beta\sqrt{t^2 - x^2/v^2}), \quad \dots\dots\dots(2)$$

thereby obtaining

$$v\int_{x/v}^{t} \xi_2(t-\mu)e^{-\alpha\mu}I_0(\beta\sqrt{\mu^2 - x^2/v^2})d\mu \Rightarrow \phi_2 e^{-\lambda x}/\lambda. \quad \dots\dots(3)$$

Thus differentiating both sides of (3) with respect to x (the validity of such procedure depending upon the function $\xi_2(t)$), we find that

$$E_x = e^{-\alpha x/v}\xi_2(t - x/v)$$

$$+ \frac{\beta x}{v}\int_{x/v}^{t} e^{-\alpha\mu}\xi_2(t-\mu)\frac{I_1(\beta\sqrt{\mu^2 - x^2/v^2})}{\sqrt{\mu^2 - x^2/v^2}}d\mu \Rightarrow \phi_2 e^{-\lambda x}, \quad \dots(4)$$

$t \geqslant x/v$, by § 13, Appendix III. E_x is the p.d. between the line and earth at any point distant x from the transmitter.

2°. *The current.* Writing ϕ_2 for E_0 in (4) § 4·31, we obtain the L.T. of the current, namely,

$$I_x \Rightarrow \phi = \sqrt{(\mathsf{C}/\mathsf{L})}\,(p + 2b)\phi_2 e^{-\lambda x}/v\lambda. \quad \dots\dots\dots(5)$$

By (3) the second term on the r.h.s. of (5) yields

$$2b\phi_2 e^{-\lambda x}/v\lambda \Subset 2b\int_{x/v}^{t} \xi_2(t-\mu)e^{-\alpha\mu}I_0(\beta\sqrt{\mu^2 - x^2/v^2})d\mu. \quad \dots\dots(6)$$

Using 22 § 8·3, with $r=0$, $n=1$, $b=x/v$, $f(0)=0$, and assuming that differentiation under the integral sign is permissible, we have for the first term on the r.h.s. of (5)

$$\frac{\partial}{\partial t}\int_{x/v}^{t} \xi_2(t-\mu)e^{-\alpha\mu}I_0(\beta\sqrt{\mu^2 - x^2/v^2})d\mu \Rightarrow p\phi_2 e^{-\lambda x}/v\lambda. \quad \dots\dots(7)$$

The l.h.s. of (7) is

$$e^{-\alpha x/v}\xi_2(t - x/v) + \int_{x/v}^{t} \xi_2(t - \mu)e^{-\alpha\mu}\left[\beta\mu\frac{I_1(\beta w)}{w} - \alpha I_0(\beta w)\right]d\mu, \quad \ldots(8)*$$

where $w = \sqrt{\mu^2 - x^2/v^2}$. Adding the r.h.s. of (6) to (8), remembering that $(2b - \alpha) = -\beta$, and incorporating the factor $\sqrt{(C/L)}$ from (5), the current at any point distant x from the transmitter is given by

$$I_x = \sqrt{(C/L)}\left\{e^{-\alpha x/v}\xi_2(t - x/v)\right.$$
$$\left. + \beta\int_{x/v}^{t} \xi_2(t - \mu)e^{-\alpha\mu}\left[\mu\frac{I_1(\beta w)}{w} - I_0(\beta w)\right]d\mu\right\}, \quad \ldots\ldots(9)$$

$t \geqslant x/v$. For the distortionless line $\beta = 0$, so (4) degenerates to

$$E_x = e^{-\alpha x/v}\xi_2(t - x/v), \quad \ldots\ldots\ldots\ldots\ldots\ldots(10)$$

and (9) degenerates to

$$I_x = \sqrt{(C/L)}\,e^{-\alpha x/v}\xi_2(t - x/v). \quad \ldots\ldots\ldots\ldots(11)$$

Thus the p.d. and current at any point of the line are in step and identical in form with the applied p.d. They bear a constant ratio to each other, namely, $\sqrt{(L/C)}$, which is the characteristic or surge impedance of the line. Both E and I are attenuated exponentially *en route*, and lag behind the p.d. at $x = 0$ by a time $t_1 = x/v$.

*4·34. Initial conditions as arbitrary functions of x.

Suppose that when $t \to -0$, the distribution of p.d. and current in the transmission line of § 4·21 are, respectively, $E = \zeta(x)$, $I = \chi(x)$, $0 \leqslant x \leqslant l$, l being the length of the line which may have any value.

Applying (5) § 2·14 to (1), (2) § 4·21, the corresponding transform equations are :

$$\frac{d\varphi}{dx} + Z\phi = pL\chi, \quad \ldots\ldots\ldots\ldots\ldots\ldots(1)$$

$$\frac{d\phi}{dx} + Y\varphi = pC\zeta. \quad \ldots\ldots\ldots\ldots\ldots\ldots(2)$$

$$*\ \frac{\partial}{\partial t}\int_{x/v}^{t} f_1(t - \mu)f_2(\beta\sqrt{\mu^2 - x^2/v^2})\,d\mu$$
$$= f_1(t - x/v)\left[f_2(\beta\sqrt{\mu^2 - x^2/v^2})\right]_{\mu = x/v} + \int_{x/v}^{t} f_1(t - \mu)\frac{\partial}{\partial\mu}(f_2)\,d\mu.$$

Assuming that the relationship corresponding to (8) § 4·21 holds, we differentiate (2) with respect to x and substitute from (1) for $d\varphi/dx$, thereby obtaining

$$\frac{d^2\phi}{dx^2} - \lambda^2\phi = g(p, x), \quad\dots\dots\dots\dots\dots(3)$$

where $\lambda^2 = \mathbf{YZ}$, and $g(p, x) = p\left(\mathbf{C}\dfrac{d\zeta}{dx} - \mathbf{LY}\chi\right)$.

The complementary function of (3) is

$$\phi_c = A_1 e^{-\lambda x} + B_1 e^{\lambda x} = A \sinh \lambda x + B \cosh \lambda x. \quad\dots\dots\dots(4)$$

The particular integral of (3) may be found by the Boole standard operational procedure as follows :

$$\phi_p = \frac{1}{D^2 - \lambda^2} \cdot g(p, x) = \frac{1}{2\lambda}\left[\frac{1}{D-\lambda} - \frac{1}{D+\lambda}\right]g(p, x), \quad\dots\dots(5)$$

$$= \frac{1}{2\lambda}\left\{e^{\lambda x}\int_0^x e^{-\lambda u}g(p, u)\,du - e^{-\lambda x}\int_0^x e^{\lambda u}g(p, u)\,du\right\}, \quad\dots(6)$$

$\lambda > 0$.

The complete solution of (3) is the sum of (4) and (6), i.e.

$$\phi = \phi_c + \phi_p. \quad\dots\dots\dots\dots\dots\dots\dots(7)$$

By (2) we have

$$-\varphi = \frac{1}{\mathbf{Y}}\frac{d\phi}{dx} - \frac{p\mathbf{C}\zeta}{\mathbf{Y}}. \quad\dots\dots\dots\dots\dots\dots(8)$$

Thus from (4), (6), and (7)

$$-\varphi = \frac{\lambda}{\mathbf{Y}}[A \cosh \lambda x + B \sinh \lambda x] + \frac{1}{\mathbf{Y}}\frac{d\phi_p}{dx} - \frac{p\mathbf{C}\zeta(x)}{\mathbf{Y}}. \quad\dots\dots(9)$$

Using the boundary conditions of § 4·331 in (9), i.e. $\varphi = \phi_1$, $x = l$, $t > 0$; $\varphi_0 = \phi_2$, $x = 0$, $t > 0$, we obtain two equations for the determination of A and B, namely,

$$-\phi_1 = \frac{\lambda}{\mathbf{Y}}(A \cosh \lambda l + B \sinh \lambda l) + \frac{1}{\mathbf{Y}}\left(\frac{d\phi_p}{dx}\right)_{x=l} - \frac{p\mathbf{C}}{\mathbf{Y}}\zeta(l), \quad\dots(10)$$

$$-\phi_2 = \frac{\lambda A}{\mathbf{Y}} + \frac{1}{\mathbf{Y}}\left(\frac{d\phi_p}{dx}\right)_{x=0} - \frac{p\mathbf{C}}{\mathbf{Y}}\zeta(0). \quad\dots\dots\dots\dots(11)$$

Having found A and B from (10), (11) the transform solutions (8), (7) for φ and ϕ are determined completely. The final step is the inversion of φ, ϕ in terms of t, and it may be necessary to apply the Mellin inversion theorem in Appendix IV.

★4·35. Particular integral of (3) § 4·34. This is given at (6) § 4·34, and may be expressed in the alternative form

$$\phi_p = \frac{1}{2\lambda}\left\{\int_0^x [e^{\lambda(x-u)} - e^{-\lambda(x-u)}] g(p, u)\, du\right\}, \quad \ldots\ldots\ldots(1)$$

$$= \frac{1}{\lambda}\int_0^x \sinh \lambda(x-u) g(p, u)\, du. \quad \ldots\ldots\ldots\ldots(2)$$

The result (2) may also be obtained using the product theorem § 2·24. In (3) § 4·34 write q for d/dx, $\phi(x) \Rightarrow \psi(q)$, $g(x) \Rightarrow \psi_1(q)$, and we get *

$$\psi = \left(\frac{q}{q^2 - \lambda^2}\right)\frac{\psi_1(q)}{q}. \quad \ldots\ldots\ldots\ldots(3)$$

Applying § 2·24 gives (2) immediately.

4·36. An analogous problem. The electrical transmission line has been treated at some length, since it is of practical importance and mathematical interest. Owing to simplification of the conditions, the solution is not exact in a physical sense.†
The problem of the transmission of longitudinal (sound) waves in a uniform homogeneous straight rod is analogous, except for lateral expansion and contraction, to that of the transmission line with **G** = 0.

* ϕ and g are functions of p and x. The L.T. is with respect to x.

† The actual conditions are complicated. An exact solution would necessitate the solution of Maxwell's equations using the appropriate boundary conditions, taking into account end effects, proximity of the earth, radiation, etc.

V

EVALUATION OF INTEGRALS AND ESTABLISHMENT OF MATHEMATICAL RELATIONSHIPS

5·11. The theorems in Chapter II, and the list of forms, can be used to great advantage in evaluating certain types of finite and infinite integrals. They are beneficial also in obtaining expansions and mathematical relationships between different functions. Integrals may be evaluated by employing one of the following procedures :

    (1) Application of § 1·22 ;

    (2)   ,,   ,, § 2·22, the infinite integral theorem :

    (3)   ,,   ,, § 2·24, the product theorem ;

    (4)   ,,   ,, formula obtained using § 2·25 ;

    (5) Introduction of a variable parameter upon which the L.T. is based. As in other chapters, the analysis depends in part upon Lerch's theorem §§ 1·14, 1·16.

*5·12. Example illustrating application of § 1·22.

Evaluate $I = \int_0^\infty \cos at \, dt / \sqrt{t}$, a real and > 0.

By 2° (b) § 5, Appendix III, the integral converges, and, therefore, the following integral converges also :

$$\int_0^\infty e^{-pt} \cos at \, dt / \sqrt{t} = \sqrt{\frac{\pi a}{2}} \int_0^\infty e^{-pt} J_{-\frac{1}{4}}(at) dt, \quad \ldots\ldots\ldots\ldots(1)$$

$$= \sqrt{\pi/2} \, (p + \sqrt{p^2 + a^2})^{\frac{1}{2}} / \sqrt{p^2 + a^2}, \quad \ldots\ldots(2)$$

from the list. Applying § 1·22

$$I = \sqrt{\pi/2} \lim_{p \to +0} (p + \sqrt{p^2 + a^2})^{\frac{1}{2}} / \sqrt{p^2 + a^2} = \sqrt{\pi/2a}, \quad \ldots\ldots\ldots(3)$$

since by 2° § 8, Appendix III, the second integral in (1) converges uniformly with respect to p in any interval $0 \leqslant p \leqslant p_0$.

*5·13. Example illustrating application of § 2·22.

Evaluate

$$\int_0^\infty J_0(a \sinh \theta) \tanh \theta \, d\theta, \quad a \text{ real} > 0.$$

Writing $\cosh \theta = t$, $\sinh \theta = \sqrt{t^2 - 1}$, $\tanh \theta \, d\theta = dt/t$ in the integral, we get

$$I = \int_1^\infty J_0(a\sqrt{t^2 - 1}) dt/t. \quad \dots\dots\dots\dots(5)$$

Using 25 § 8·3,

$$I = \int_0^\infty \frac{\phi(p)}{p} dp = \int_0^\infty e^{-\sqrt{p^2 + a^2}} dp/\sqrt{p^2 + a^2}. \quad \dots\dots\dots(6)$$

Substituting $p = a \sinh \varphi$ in the second integral in (6) gives

$$I = \int_0^\infty e^{-a \cosh \varphi} \, d\varphi = K_0(a), \quad \dots\dots\dots\dots(7)$$

by reference 11, p. 165. Hence if C_1 § 9, Appendix III is satisfied,

$$\int_0^\infty J_0(a \sinh \theta) \tanh \theta \, d\theta = K_0(a). \quad \dots\dots\dots(8)$$

Now (a) $f(t) = J_0(a\sqrt{t^2 - 1})$, and this is continuous in $t > 1$;

(b) $\int_1^\infty e^{-pt} J_0(a\sqrt{t^2 - 1}) dt/t$ is u.c. in $0 \leqslant p \leqslant p_2$;

(c) $\int_0^\infty e^{-pt} J_0(a\sqrt{t^2 - 1}) dp$ is u.c. in $1 \leqslant t \leqslant h_2$ and in $t > 1$, since the integrand is continuous therein; *

(d') $\int_1^h J_0(a\sqrt{t^2 - 1}) dt/t$ exists, since the integrand is continuous in $1 < t \leqslant h$, also

$$\int_h^\infty \frac{|J_0(a\sqrt{t^2 - 1})|}{t} dt \sim \int_h^\infty \sqrt{(2/\pi a)} |\cos(a\sqrt{(t^2 - 1)} - \tfrac{1}{4}\pi)| \frac{dt}{t^{3/2}}$$

when h is large enough, and this is $< \sqrt{\dfrac{2}{\pi a}} \int_h^\infty dt/t^{3/2} = 2\sqrt{2}/\sqrt{\pi a h}$,

so the integral is convergent. Hence the conditions for inverting the order of integration are satisfied.

* See § 1·21.

5·141. Examples illustrating application of product theorem § 2·24.

The Beta function, due to Euler, is defined by

$$B(\mu, \nu) = \int_0^1 x^{\mu-1}(1-x)^{\nu-1}\,dx, \qquad \ldots\ldots\ldots\ldots(1)$$

and it is desired to evaluate the integral in terms of known functions. Now $t^{\mu-1} \Rightarrow \Gamma(\mu)/p^{\mu-1} = \phi_1$, $t^{\nu-1} \Rightarrow \Gamma(\nu)/p^{\nu-1} = \phi_2$ the integrals of type (1) § 1·11 being *absolutely* convergent if $R(\mu) > 0$, $R(\nu) > 0$, $R(p) > 0$, so the conditions for validity of § 2·24 are satisfied. Thus

$$\phi_1\phi_2/p = \Gamma(\mu)\Gamma(\nu)/p^{\mu+\nu-1} \subset \Gamma(\mu)\Gamma(\nu)t^{\mu+\nu-1}/\Gamma(\mu+\nu). \quad \ldots(2)$$

By § 2·24 the r.h.s. of (2) is equal to

$$\int_0^t x^{\mu-1}(t-x)^{\nu-1}\,dx. \qquad \ldots\ldots\ldots\ldots\ldots(3)$$

Writing $t = 1$ in (2), (3) we obtain

$$\int_0^1 x^{\mu-1}(1-x)^{\nu-1}\,dx = \frac{\Gamma(\mu)\Gamma(\nu)}{\Gamma(\mu+\nu)} = B(\mu, \nu), \qquad \ldots\ldots\ldots(4)$$

$R(\mu) > 0$, $R(\nu) > 0$.

By substituting $x = \sin^2\theta$ and writing $\mu + 1/2$, $\nu + 1/2$ for μ and ν, respectively, in (4), we get the well-known formula

$$\int_0^{\frac{1}{2}\pi} \sin^{2\mu}\theta \cos^{2\nu}\theta\,d\theta = \frac{\Gamma(\mu+1/2)\Gamma(\nu+1/2)}{2\Gamma(\mu+\nu+1)}, \qquad \ldots\ldots\ldots(5)$$

$$= \tfrac{1}{2}B(\mu+\tfrac{1}{2}, \nu+\tfrac{1}{2}), \qquad \ldots\ldots\ldots\ldots(6)$$

$R(\mu) > -\tfrac{1}{2}$, $R(\nu) > -\tfrac{1}{2}$.

5·142. Evaluate $I = \int_0^t J_\mu(\lambda)J_\nu(t-\lambda)\,d\lambda$.

By 4° § 1·211, $\int_0^\infty e^{-pt}J_\mu(t)\,dt$ is *absolutely* convergent if $R(\mu) > -1$, $R(\nu) > -1$, so § 2·24 is valid in the present case. Now

$$J_\mu(t) \Rightarrow \frac{p}{\sqrt{p^2+1}}(\sqrt{p^2+1}-p)^\mu = \frac{pb^\mu}{\sqrt{p^2+1}}, \qquad \ldots\ldots\ldots\ldots(1)$$

where $b = \sqrt{p^2+1} - p$; also

$$2b/(1+b^2) = 2b/2(p^2+1-p\sqrt{p^2+1})$$

$$= 2b/2b\sqrt{p^2+1} = 1/\sqrt{p^2+1}. \qquad \ldots\ldots\ldots\ldots(2)$$

Applying § 2·24, we have

$$I = \phi_1\phi_2/p = pb^{\mu+\nu}/(p^2+1) = \frac{2pb^{\mu+\nu}}{\sqrt{p^2+1}} \cdot \frac{b}{1+b^2}, \quad \dots\dots(3)$$

$$= \frac{2pb^{\mu+\nu+1}}{\sqrt{p^2+1}} \sum_{r=0}^{\infty} (-1)^r b^{2r}, \quad \dots\dots\dots\dots(4)$$

provided $|b| < 1$, i.e. $p > 0$. Interpreting (4) by means of (1), and using Lerch's theorem §§ 1·14, 1·16 we find that

$$\int_0^t J_\mu(\lambda) J_\nu(t-\lambda)d\lambda = \int_0^t J_\mu(t-\lambda) J_\nu(\lambda)d\lambda$$

$$= 2 \sum_{r=0}^{\infty} (-1)^r J_{\mu+\nu+2r+1}(t). \quad \dots\dots(5)^*$$

The conditions in § 2·18 are satisfied, so term by term interpretation in (4) and (5) is valid.

5·143. Evaluate $I = \int_0^t J_\nu(\lambda) \sin(t-\lambda)d\lambda.$

By 4° § 1·211, $\int_0^{\infty} e^{-pt}J_\nu(t)dt$ is *absolutely* convergent if $R(\nu) > -1$, and by § 1·21, since $\sin t$ is bounded in $t \geqslant 0$, $\int_0^{\infty} e^{-pt}\sin t \, dt$ is *absolutely* convergent. Thus the conditions for the validity of § 2·24 are satisfied. Then if $R(\nu) > -1$

$$J_\nu(t) = pb^\nu/\sqrt{p^2+1} = \phi_1, \quad \dots\dots\dots\dots(1)$$

and $\quad\quad \sin t = p/(p^2+1) \quad = \phi_2 \quad \dots\dots\dots\dots(2)$

b being defined in § 5·142.

Applying § 2·24, we have

$$I = \phi_1\phi_2/p = pb^\nu/(p^2+1)^{3/2} = \frac{4pb^{\nu+2}}{\sqrt{p^2+1}} \cdot \frac{1}{(1+b^2)^2}, \quad \dots\dots(3)$$

$$= \frac{4pb^{\nu+2}}{\sqrt{p^2+1}} \sum_{r=0}^{\infty} (-1)^r (r+1)b^{2r}, \quad \dots\dots\dots(4)$$

if $|b| < 1$, i.e. $p > 0$. Hence by (1) § 5·142

$$I = 4 \sum_{r=0}^{\infty} (-1)^r (r+1) J_{\nu+2r+2}(t), \quad \dots\dots\dots(5)$$

* Writing $\mu = 0$, $\nu = 1$ and using (3) § 2·242, we obtain the well-known expansion $\cos t = J_0(t) + 2 \sum_{r=1}^{\infty} (-1)^r J_{2r}(t)$, [see reference 11, p. 43].

the conditions for the term by term interpretation in (4), (5) being satisfied.

*5·144. Alternative form of (5) § 5·143 when $v=2m$, an even positive integer. We have

$$\frac{\phi_1\phi_2}{p} = \frac{pb^{2m}}{(p^2+1)^{3/2}} = \frac{pb^{2m+1}}{(p^2+1)^{3/2}}(2m\sqrt{p^2+1}+1/b) - \frac{2mpb^{2m+1}}{p^2+1}. \quad \dots(1)$$

Now

$$-\frac{2mpb^{2m+1}}{p^2+1} = (-1)^m\frac{2mp}{\sqrt{p^2+1}}\left[(-1)^{m+1}\left(\frac{2b}{1+b^2}\right)b^{2m+1}\right], \quad \dots(2)$$

$$= (-1)^m\frac{2mp}{\sqrt{p^2+1}}\left[-\frac{2b^2}{1+b^2}+2b^2\left\{\frac{1+(-1)^{m+1}b^{2m}}{1+b^2}\right\}\right], \quad \dots(3)$$

$$= (-1)^m\,2m\left[\frac{p^2}{p^2+1}-\frac{p}{\sqrt{p^2+1}}+\right.$$

$$\left.+\frac{2p}{\sqrt{p^2+1}}\{b^2-b^4+\dots+(-1)^{m+1}b^{2m}\}\right], \quad \dots(4)$$

since $-2b^2/(1+b^2) = -b/\sqrt{p^2+1} = -1+p/\sqrt{p^2+1}$.

Hence from (1) and (4)

$$\frac{\phi_1\phi_2}{p} = \frac{pb^{2m+1}}{(p^2+1)^{3/2}}\left(2m\sqrt{p^2+1}+\frac{1}{b}\right) + \text{r.h.s. of (4).} \quad \dots\dots(5)$$

Interpreting both sides of (5), we obtain for $m \geqslant 0$,

$$\int_0^t J_{2m}(\lambda)\sin(t-\lambda)d\lambda = tJ_{2m+1}(t)* + (-1)^m 2m[\cos t - J_0(t)$$

$$+ 2J_2(t) - \dots + (-1)^{m+1}\,2J_{2m}(t)],$$

$$= A. \quad \dots\dots\dots\dots\dots\dots\dots\dots\dots\dots\dots\dots\dots\dots\dots(6)$$

In like manner it can be shown that if $m>0$

$$\int_0^t J_{2m}(\lambda)\cos(t-\lambda)d\lambda = tJ_{2m-1}(t) + (-1)^{m+1}2m[\sin t - 2J_1(t)$$

$$+ 2J_3(t) - \dots + (-1)^m 2J_{2m-1}(t)],$$

$$= B. \quad \dots\dots\dots\dots\dots\dots\dots\dots\dots\dots\dots\dots\dots(7)$$

* The L.T. of this term is found by applying the r.h.s. of (3) § 2·17 to the L.T. for $J_{2m+1}(t)$ in the form at § 5·142.

The reader will now have no difficulty in demonstrating that

$$\int_0^t J_{2m}(\lambda)\sin\lambda\,d\lambda = B\sin t - A\cos t, \quad\dots\dots\dots(8)$$

and

$$\int_0^t J_{2m}(\lambda)\cos\lambda\,d\lambda = A\sin t + B\cos t. \quad\dots\dots\dots(9)$$

When $\nu=0$ in (5) § 5·143 and $m=0$ in (6), we obtain the known relationship [11]

$$tJ_1(t) = 4\sum_{r=0}^{\infty}(-1)^r(r+1)J_{2r+2}(t). \quad\dots\dots\dots(10)$$

By giving ν and m in the above formulae the values 1, 2, 3, ..., a series of identities can be written down. By evaluating the l.h.s. of (7) as an infinite series of Bessel functions, another set of identities may be obtained.

*5·145. Show that

$$I = \int_0^{\frac{1}{2}\pi} J_\mu(a\sin\theta)I_\nu(a\cos\theta)\tan^{\mu+1}\theta\,d\theta$$

$$= \frac{\Gamma[\frac{1}{2}(\nu-\mu)]}{2\Gamma[\frac{1}{2}(\nu+\mu)+1]}\left(\frac{a}{2}\right)^\mu J_\nu(a), \ a \text{ real}>0 \text{ [ref. 5].}$$

Write $a = 2\sqrt{t}$, $\cos\theta = \sqrt{\lambda/t}$, which gives

$$\sin\theta = \sqrt{(t-\lambda)/t}, \ \tan\theta = \sqrt{(t-\lambda)/\lambda}, \ d\theta = -d\lambda/2\sqrt{\lambda(t-\lambda)}.$$

These substitutions lead to

$$I = \tfrac{1}{2}\int_0^t(t-\lambda)^{\frac{1}{2}\mu}J_\mu(2\sqrt{t-\lambda}) \ . \ \lambda^{-(1+\frac{1}{2}\mu)}I_\nu(2\sqrt{\lambda})d\lambda, \quad\dots\dots(1)$$

so in the symbolism of § 2·24

$$f_1(t) = t^{\frac{1}{2}\mu}J_\mu(2\sqrt{t}), \ f_2(t) = t^{-(1+\frac{1}{2}\mu)}I_\nu(2\sqrt{t}).$$

Following the procedure in 4° §1·211 it may be demonstrated that

$$\int_0^\infty e^{-pt}f_1(t)dt \text{ is } absolutely \text{ convergent if } R(\mu)>-1,$$

and that

$$\int_0^\infty e^{-pt}f_2(t)dt \text{ is } absolutely \text{ convergent if } R(\nu-\mu)>0,$$

i.e. $R(\nu)>R(\mu)$.

With these restrictions on μ and ν, we may apply § 2·24. Then from the list

$$t^{\frac{1}{2}\mu}J_{\mu}(2\sqrt{t}) \Rightarrow p^{-\mu}e^{-1/p} = \phi_1, \quad \dots\dots\dots(2)$$

and

$$t^{-(1+\frac{1}{2}\mu)}I_{\nu}(2\sqrt{t})$$

$$\Rightarrow \frac{\Gamma[\frac{1}{2}(\nu-\mu)]}{\Gamma(\nu+1)} p^{\frac{1}{2}(\mu-\nu)+1}{}_1F_1[\frac{1}{2}(\nu-\mu)\,;\,\nu+1\,;\,1/p] = \phi_2, \quad\dots(3)$$

where $\quad {}_1F_1(\alpha\,;\,\beta\,;\,\gamma) = 1 + \dfrac{\alpha}{1!\beta}\gamma + \dfrac{\alpha(\alpha+1)}{2!\beta(\beta+1)}\gamma^2 + \dots\,.$

By (2), (3)

$$2I \Rightarrow \phi_1\phi_2/p$$

$$= \frac{\Gamma[\frac{1}{2}(\nu-\mu)]}{\Gamma(\nu+1)} p^{-\frac{1}{2}(\nu+\mu)}e^{-1/p}{}_1F_1[\frac{1}{2}(\nu-\mu)\,;\,\nu+1\,;\,1/p]. \quad\dots(4)$$

Applying the formula

$${}_1F_1(\alpha\,;\,\beta\,;\,\gamma) = e^{\gamma}{}_1F_1[(\beta-\alpha)\,;\,\beta\,;\,-\gamma],$$

(4) may be written

$$2I \Rightarrow \frac{\Gamma[\frac{1}{2}(\nu-\mu)]}{\Gamma(\nu+1)} p^{-\frac{1}{2}(\nu+\mu)}{}_1F_1[\frac{1}{2}(\nu+\mu)+1\,;\,\nu+1\,;\,-1/p]. \quad\dots(5)$$

From the list

$$t^{\frac{1}{2}\mu}J_{\nu}(2\sqrt{t})$$

$$\Rightarrow \frac{\Gamma[\frac{1}{2}(\nu+\mu)+1]}{\Gamma(\nu+1)} p^{-\frac{1}{2}(\nu+\mu)}{}_1F_1[\frac{1}{2}(\nu+\mu)+1\,;\,\nu+1\,;\,-1/p], \quad\dots(6)$$

so by (5) and (6) we find that

$$I = \frac{\Gamma[\frac{1}{2}(\nu-\mu)]t^{\frac{1}{2}\mu}}{2\Gamma[\frac{1}{2}(\nu+\mu)+1]} J_{\nu}(2\sqrt{t}), \quad\dots\dots\dots(7)$$

$$= \frac{\Gamma[\frac{1}{2}(\nu-\mu)]}{2\Gamma[\frac{1}{2}(\nu+\mu)+1]} \left(\frac{a}{2}\right)^{\mu} J_{\nu}(a), \quad\dots\dots(8)$$

$R(\nu) > R(\mu) > -1.$

*5·151. Examples illustrating application of formulae obtained using § 2·25.

For a rigorous application of this complicated theorem, it is necessary, with a given $f(t)$, to prove 1°, that integral (1) § 2·25 converges ; 2°, that the inverted order of integration in (4) § 2·25 is permissible. These proofs are often long, and may necessitate reference to analysis in other works.   Here we shall give purely formal illustrations of theorem 14.   We may remark, however, that if integrals (1), (5) § 2·25 converge, the results are usually correct.

Application of $\int_0^\infty J_0(2\sqrt{xt})f(x)\,dx = p\,\phi(1/p)$.

(a) Take $f(x) = \sin x$, then $\phi(p) = p/(p^2 + 1)$, so

$$p\,\phi(1/p) = p^2/(p^2 + 1) \Subset \cos t. \quad\ldots\ldots\ldots\ldots(1)$$

Hence $$\int_0^\infty J_0(2\sqrt{xt})\sin x\,dx = \cos t. \quad\ldots\ldots\ldots\ldots(2)$$

The convergence of this integral is examined in 5° § 5, Appendix III.

(b) Take $f(x) = e^{-x}x^n$, then $\phi(p) = n!p/(p+1)^{n+1}$, $\ldots\ldots\ldots\ldots$(3)

n being a positive integer.   Now

$$p\,\phi(1/p) = n!p^{n+1}/(p+1)^{n+1} \Subset n!e^{-t}L_n(t), \quad\ldots\ldots\ldots\ldots(4)$$

by the list.   Hence

$$\int_0^\infty J_0(2\sqrt{xt})e^{-x}x^n\,dx = n!e^{-t}L_n(t), \quad\ldots\ldots\ldots\ldots(5)$$

$L_n(t)$ being the Laguerre polynomial of order n [13].   The reader should investigate the convergence of this integral.

*5·152. Example.

We apply the theorem $\dfrac{1}{\sqrt{\pi t}}\displaystyle\int_0^\infty e^{-x^2/4t}f(x)\,dx = \phi(p^{1/2})$ in two cases.

(a) Take $f(x) = \sinh x$, then $\phi(p) = p/(p^2 - 1)$, so

$$\phi(p^{1/2}) = p^{1/2}/(p-1) \Subset e^t\,\mathrm{erf}\,t^{1/2}, \ldots\ldots\ldots\ldots(1)$$

from the list.   Hence if $t > 0$,

$$\int_0^\infty e^{-x^2/4t}\sinh x\,dx = \sqrt{\pi t}\,e^t\,\mathrm{erf}\,t^{1/2}. \quad\ldots\ldots\ldots\ldots(2)$$

(b) Take $f(x) = \mathrm{bei}'x$, then $\mathrm{bei}\,(0) = 0$, and

$$\phi(p) = \frac{p^2(\sqrt{p^4 + 1} - p^2)^{1/2}}{\sqrt{2}\,(p^4 + 1)},$$

so

$$\phi(p^{1/2}) = \frac{p(\sqrt{p^2+1}-p)^{1/2}}{\sqrt{2}(p^2+1)} = \frac{p/\sqrt{2}}{\sqrt{p^2+1}(p+\sqrt{p^2+1})^{1/2}}, \quad ...(3)$$

$$= J_{\frac{1}{4}}(t)/\sqrt{2}. \quad ....(4)$$

Hence if $t > 0$,

$$\int_0^\infty e^{-x^2/4t} \, \mathrm{bei}'x \, dx = \sqrt{\pi t/2} \, J_{\frac{1}{4}}(t) = \sin t. \quad ....(5)$$

*5·153. Application of

$$t^\nu \int_0^\infty J_{2\nu}(2\sqrt{xt}) f(x) x^{-\nu} \, dx = p^{1-2\nu} \phi(1/p), \quad R(\nu) > -\tfrac{1}{2}.$$

Take $f(x) = e^{-ax}J_\nu(x)x^\nu$, a real > 0, then

$$\phi(p) = \frac{2^\nu}{\sqrt{\pi}} \frac{\Gamma(\nu+\tfrac{1}{2})p}{[(p+a)^2+1]^{\nu+1/2}},$$

and

$$p^{1-2\nu}\phi(1/p) = \frac{2^\nu}{\sqrt{\pi}} \frac{\Gamma(\nu+\tfrac{1}{2})p}{[p^2+(1+ap)^2]^{\nu+1/2}}, \quad ....(1)$$

Now

$$p^2+(1+ap)^2 = p^2(a^2+1)+2ap+1, \quad ....(2)$$

$$= (a^2+1)\left[p^2+\frac{2ap}{a^2+1}+\left(\frac{a}{a^2+1}\right)^2+\frac{1}{(a^2+1)^2}\right], \quad ...(3)$$

$$= \left[\left(p+\frac{a}{a^2+1}\right)^2+\frac{1}{(a^2+1)^2}\right](a^2+1). \quad ....(4)$$

Substituting from (4) into (1), and interpreting in terms of t by aid of the formula above (1) and (4) § 2·11, we obtain

$$\frac{2^\nu\Gamma(\nu+\tfrac{1}{2})}{\sqrt{\pi}} \cdot \frac{p}{\left[\left(p+\dfrac{a}{a^2+1}\right)^2+\left(\dfrac{1}{a^2+1}\right)^2\right]^{\nu+1/2}} \cdot \frac{1}{(a^2+1)^{\nu+1/2}}$$

$$= \frac{e^{-at/(a^2+1)} \, t^\nu}{(a^2+1)^{1/2}} J_\nu[t/(a^2+1)]. \quad ......(5)$$

Hence

$$\int_0^\infty e^{-ax}J_{2\nu}(2\sqrt{xt})J_\nu(x)\,dx = \frac{e^{-at/(a^2+1)}}{(a^2+1)^{1/2}} J_\nu[t/(a^2+1)], ...(6)$$

$R(\nu) > -\frac{1}{2}$, or

$$\int_0^\infty e^{-ax}J_{2\nu}(2t\sqrt{x})\,J_\nu(x)\,dx = \frac{e^{-at^2/(a^2+1)}}{(a^2+1)^{1/2}}\,J_\nu[t^2/(a^2+1)].\quad\ldots(7)$$

Writing p for a, b for t, and t for x, it follows from (7) that

$$J_{2\nu}(2b\sqrt{t})J_\nu(t) \rightleftharpoons \frac{pe^{-pb^2/(p^2+1)}}{\sqrt{p^2+1}}\,J_\nu[b^2/(p^2+1)],\quad\ldots\ldots(8)$$

$R(\nu) > -\frac{1}{2}$, b real >0.

*5·154. Example. As a second illustration of the application of the theorem in § 5·153, we write $\frac{1}{2}\nu$ for ν and take

$$f(x) = e^{-ax}x^{\frac{1}{2}\nu}J_\nu(2\sqrt{x}).$$

Then from the list

$$x^{\frac{1}{2}\nu}J_\nu(2\sqrt{x}) \rightleftharpoons p^{-\nu}e^{-1/p},\quad\ldots\ldots\ldots\ldots(1)$$

$R(\nu) > -1$. Applying § 2·13 to (1), with $a > 0$, we obtain

$$f(x) = e^{-ax}x^{\frac{1}{2}\nu}J_\nu(2\sqrt{x}) \rightleftharpoons pe^{-1/(p+a)}/(p+a)^{\nu+1},\quad\ldots\ldots(2)$$

so

$$p^{1-\nu}\phi(1/p) = pe^{-p/(pa+1)}/(pa+1)^{\nu+1},\quad\ldots\ldots\ldots(3)$$

$$= \frac{e^{-1/a}}{a^{\nu+1}}\cdot\frac{p}{(p+1/a)^{\nu+1}}\cdot e^{1/a^2(p+1/a)}.\quad\ldots\ldots(4)$$

$$\rightleftharpoons e^{-(t+1)/a}a^{-1}t^{\frac{1}{2}\nu}I_\nu(2\sqrt{t/a}),\quad\ldots\ldots\ldots(5)$$

by aid of the list and § 2·13. Hence by § 5·153, (2) and (5) above, we have

$$a\int_0^\infty e^{-ax}J_\nu(2\sqrt{xt})\,J_\nu(2\sqrt{x})\,dx = e^{-(t+1)/a}I_\nu(2\sqrt{t/a}),\quad\ldots\ldots(6)$$

with $R(\nu) > -1$ for convergence at the lower limit (see 2° § 8, Appendix III). Substituting p for a, t for x, and a for t in (6) we obtain

$$J_\nu(2\sqrt{at})J_\nu(2\sqrt{t}) \rightleftharpoons e^{-(a+1)/p}I_\nu(2\sqrt{a/p}),\quad\ldots\ldots\ldots(7)$$

$R(\nu) > -1$. Writing $2\sqrt{x} = v$, $dx = v\,dv/2$, and t^2 for t in (6) leads to

$$a\int_0^\infty e^{-av^2/4}J_\nu(vt)\,J_\nu(v)\,v\,dv = 2e^{-(t^2+1)/a}I_\nu(2t/a),\quad\ldots\ldots(8)$$

$R(\nu) > -1$, which is Weber's second exponential integral.

5·161. Example illustrating introduction of variable parameter.

Evaluate
$$I = \int_0^\infty J_\nu(x)\,dx/x^{\nu-\mu+1}.$$

Let $x = 2\sqrt{at}$, a real > 0, then $dx = \sqrt{a/t}\,dt$, and the integral assumes the form

$$2^{\nu-\mu+1}a^{\frac{1}{2}(\nu-\mu)}I = \int_0^\infty J_\nu(2\sqrt{at})\,dt/t^{\frac{1}{2}(\nu-\mu+2)}. \qquad\ldots\ldots\ldots(1)$$

Multiplying both sides of (1) by $a^{\frac{1}{2}\nu}$, we get

$$2^{\nu-\mu+1}a^{\nu-\frac{1}{2}\mu}I = \int_0^\infty a^{\frac{1}{2}\nu}J_\nu(2\sqrt{at})\,dt/t^{\frac{1}{2}(\nu-\mu+2)}. \qquad\ldots\ldots\ldots(2)$$

Taking $a = 1/p$, we find from the list that

$$2^{\nu-\mu+1}a^{\nu-\frac{1}{2}\mu}I = p^{-\nu}\int_0^\infty e^{-t/p}\,t^{\frac{1}{2}\mu-1}\,dt, \qquad\ldots\ldots\ldots\ldots(3)$$

$$= \Gamma(\tfrac{1}{2}\mu)p^{\frac{1}{2}\mu-\nu}, \qquad\ldots\ldots\ldots\ldots\ldots(4)$$

provided $R(\mu) > 0$ and the order of integration, in the repeated integral for the L.T. of (2), is invertible.* The inversion can be proved valid, so that inverting the r.h.s. of (4) in terms of a leads to the result

$$2^{\nu-\mu+1}a^{\nu-\frac{1}{2}\mu}I = \frac{\Gamma(\tfrac{1}{2}\mu)a^{\nu-\frac{1}{2}\mu}}{\Gamma(\nu-\tfrac{1}{2}\mu+1)}, \qquad\ldots\ldots\ldots\ldots(5)$$

or
$$I = \frac{\Gamma(\tfrac{1}{2}\mu)}{2^{\nu-\mu+1}\Gamma(\nu-\tfrac{1}{2}\mu+1)}. \qquad\ldots\ldots\ldots(6)$$

This result is known as Weber's integral.

---

* Multiplying the r.h.s. of (2) by ae^{-pa} and integrating with respect to a from 0 to $+\infty$, we get

$$2^{\nu-\mu+1}a^{\nu-\frac{1}{2}\mu}I = a\int_0^\infty e^{-pa}\left[\int_0^\infty a^{\frac{1}{2}\nu}J_\nu(2\sqrt{at})\,dt/t^{\frac{1}{2}(\nu-\mu+2)}\right]da.$$

Using § 9B, Appendix III, it may be shown that the order of integration is invertible, so the foregoing may be written

$$2^{\nu-\mu+1}a^{\nu-\frac{1}{2}\mu}I = a\int_0^\infty t^{\frac{1}{2}\mu-1}\,dt\int_0^\infty e^{-pa}(a/t)^{\frac{1}{2}\nu}J_\nu(2\sqrt{at})\,da,$$

$$= p^{-\nu}\int_0^\infty e^{-t/p}t^{\frac{1}{2}\mu-1}\,dt \quad\text{as in (3).}$$

5·162. Example. Evaluate $I = \int_0^\infty t \sin at \, dt/(1+t^2)$, a real > 0.

Taking $a \Rightarrow 1/p$ and replacing $\sin at$ by its L.T., we get

$$\int_0^\infty \frac{t \sin at}{1+t^2} \, dt \Rightarrow \int_0^\infty \frac{t^2 p \, dt}{(1+t^2)(p^2+t^2)}, \quad \dots\dots\dots(1)$$

provided that the order of integration in the repeated integral corresponding to the r.h.s. of (1) may be inverted. This step can be proved valid,* so we resolve (1) into partial fractions thereby obtaining

$$I = \left(\frac{p}{p^2-1}\right) \int_0^\infty \left(\frac{p^2}{t^2+p^2} - \frac{1}{t^2+1}\right) dt, \quad \dots\dots\dots(2)$$

$$= \left(\frac{p}{p^2-1}\right)\left[\frac{\pi}{2}p - \frac{\pi}{2}\right], \quad \dots\dots\dots\dots(3)$$

$$= \tfrac{1}{2}\pi \frac{p}{p+1} \subset \tfrac{1}{2}\pi e^{-a}. \quad \dots\dots\dots\dots(4)$$

Hence $\qquad \int_0^\infty \frac{t \sin at}{1+t^2} \, dt = \tfrac{1}{2}\pi e^{-a}. \quad \dots\dots\dots\dots(5)$

5·21. Expansion of $- \mathrm{Ei}(-t) = \int_t^\infty e^{-x} \, dx/x, \; 0 < t < \infty$.

From the list we find that †

$$- Ei(-t) \Rightarrow \log(p+1) = \log p(1+1/p) = \log p + \log(1+1/p), \dots(1)$$

$$= \log p + \frac{1}{p} - \frac{1^{\cdot}}{2p^2} + \frac{1}{3p^3} - \dots, \quad \dots\dots\dots\dots(2)$$

the series being absolutely convergent if $|p| > 1$, a condition which can be satisfied. Inverting (2) term by term by aid of the list, we get

$$- Ei(-t) = -\gamma - \log t + t - \frac{t^2}{2 \cdot 2!} + \frac{t^3}{3 \cdot 3!} - \dots, \quad \dots(3)$$

$$= -\gamma - \log t - \sum_{r=1}^\infty (-1)^r t^r/r \cdot r! \quad \dots\dots\dots(4)$$

* Using analysis akin to that in 2° § 10, Appendix III.

† The L.T. of $- Ei(-t)$ is found by applying theorem 12, § 2·23. Then since $e^{-t} \Rightarrow p/(p+1)$, we have $\int_t^\infty e^{-x} \, dx/x \Rightarrow \int_0^p dp/(p+1) = \log(p+1)$,

The *series* in (4) is uniformly convergent in any closed interval $0 \leqslant t \leqslant h$, so the conditions in § 2·18 are satisfied, and term by term correspondence is valid.

*5·22. Expansion of $J_\nu(2a\sqrt{t})J_\nu(2b\sqrt{t})$ in terms of Bessel functions.

From the list

$$J_\nu(2a\sqrt{t}) J_\nu(2b\sqrt{t}) \rightleftharpoons e^{-(a^2+b^2)/p} I_\nu(2ab/p), \quad \ldots\ldots\ldots(1)$$

$R(\nu) > -1$, a and b real > 0. Now

$$e^{-(a^2+b^2)/p} I_\nu(2ab/p) = e^{-(a^2+b^2)/p} \sum_{r=0}^{\infty} (ab/p)^{\nu+2r}/r!\,\Gamma(\nu+r+1), \quad \ldots(2)$$

whilst

$$t^{\nu/2} J_\nu(2c\sqrt{t}) \rightleftharpoons c^\nu p^{-\nu} e^{-c^2/p}. \quad \ldots\ldots\ldots\ldots\ldots(3)$$

Interpreting the r.h.s. of (2) term by term using (3), we obtain

$$J_\nu(2a\sqrt{t}) J_\nu(2b\sqrt{t}) = \sum_{r=0}^{\infty} \left(\frac{a^2 b^2 t}{a^2+b^2}\right)^{\frac{\nu}{2}} \frac{J_{\nu+2r}[2\sqrt{t(a^2+b^2)}]}{r!\,\Gamma(\nu+r+1)}, \quad \ldots(4)$$

the term by term correspondence being valid, since the conditions in § 2·18 are satisfied. Writing $\tfrac{1}{2}a$ for both a, b, and t for $\sqrt{t}$ in (4), we find that

$$J_\nu^2(at) = \sum_{r=0}^{\infty} \left(\frac{at}{2\sqrt{2}}\right)^{\nu+2r} \frac{J_{\nu+2r}(at\sqrt{2})}{r!\,\Gamma(\nu+r+1)}. \quad \ldots\ldots\ldots\ldots(5)$$

*5·23. Expansion of $pe^{-b\sqrt{p^2+a^2}}/\sqrt{p^2+a^2}$ in terms of $K_\mu(bp)$.

From the list

$$(t^2 - b^2)^\mu \rightleftharpoons \frac{(2b)^{\mu+1/2} \Gamma(\mu+1)}{\sqrt{\pi}\, p^{\mu-1/2}} K_{\mu+\frac{1}{2}}(bp), \quad \ldots\ldots\ldots(1)$$

$t > b$, b real, $R(\mu) \geqslant -\tfrac{1}{2}$. Now

$$J_\nu(a\sqrt{t^2-b^2}) = \sum_{r=0}^{\infty} (-1)^r \frac{(\tfrac{1}{2}a)^{\nu+2r}(t^2-b^2)^{\frac{1}{2}\nu+r}}{r!\,\Gamma(\nu+r+1)}, \quad \ldots\ldots\ldots(2)$$

so it follows from (1) that

$J_\nu(a\sqrt{t^2-b^2})$

$$\rightleftharpoons \frac{1}{\sqrt{\pi}} \sum_{r=0}^{\infty} (-1)^r \frac{a^{\nu+2r} b^{\frac{1}{2}(\nu+1)+r} \Gamma(\tfrac{1}{2}\nu+r+1)}{r!\,\Gamma(\nu+r+1)(2p)^{\frac{1}{2}\nu+r-1/2}} K_{\frac{1}{2}\nu+r+\frac{1}{2}}(pb), \quad \ldots(3)$$

$R(\nu) > -1$. The series in (2) is absolutely and uniformly convergent in $b \leqslant t \leqslant h$, whilst series (3) converges absolutely, thereby satisfying (a), (b) in § 2·18. Furthermore

$$J_0(a\sqrt{t^2-b^2}) \rightleftharpoons pe^{-b\sqrt{p^2+a^2}}/\sqrt{p^2+a^2}, \quad \ldots\ldots\ldots\ldots(4)$$

so by Lerch's theorem §§ 1·14, 1·16, if $\nu = 0$ in (3), we obtain the expansion

$$\frac{pe^{-b\sqrt{p^2+a^2}}}{\sqrt{p^2+a^2}} = \frac{1}{\pi^{1/2}} \sum_{r=0}^{\infty} (-1)^r \frac{a^{2r}b^{r+1/2}}{r! \, (2p)^{r-1/2}} K_{\frac{1}{2}+r}(pb). \quad \ldots\ldots\ldots(5)$$

*5·24. Expansion of $J_\nu(u)$ in terms of Laguerre polynomials.

From the list, when a is real and > 0, $R(\nu) > -1$,

$$t^\nu a^n L_n{}^\nu(t) \Rightarrow \frac{a^n \Gamma(\nu+n+1)}{n! \, p^\nu} (1 - 1/p)^n, \quad \ldots\ldots\ldots\ldots(1)$$

so

$$t^\nu \sum_{n=0}^{\infty} \frac{a^n L_n{}^\nu(t)}{\Gamma(\nu+n+1)} \Rightarrow \frac{e^{a-a/p}}{p^\nu} \subset (t/a)^{\frac{1}{2}\nu} e^a J_\nu(2\sqrt{at}), \quad \ldots\ldots\ldots(2)$$

or

$$J_\nu(2\sqrt{at}) = e^{-a}(at)^{\nu/2} \sum_{n=0}^{\infty} \frac{a^n L_n{}^\nu(t)}{\Gamma(\nu+n+1)}. \quad \ldots\ldots\ldots\ldots(3)$$

Writing $a = 1$ and $t = \frac{1}{4}u^2$ gives, with $R(\nu) > -1$,

$$J_\nu(u) = e^{-1}(\tfrac{1}{2}u)^\nu \sum_{n=0}^{\infty} \frac{L_n{}^\nu(\frac{1}{4}u^2)}{\Gamma(\nu+n+1)}. \quad \ldots\ldots\ldots\ldots(4)$$

The analysis is valid, since the conditions in § 2·18 are satisfied. This should be confirmed by the reader.

*5·25. Expansion of $\cos(t - \frac{1}{4}\pi)$ in terms of Laguerre polynomials.

From the list and (4) § 2·11

$$L_{2n}(2t) \Rightarrow (1 - 2/p)^{2n}. \quad \ldots\ldots\ldots\ldots\ldots\ldots(1)$$

By (2) § 2·13 and (1) above

$$e^{-t}L_{2n}(2t) \Rightarrow \frac{p}{p+1}\left(\frac{p-1}{p+1}\right)^{2n} = \frac{p}{p+1}\beta^{2n}, \quad \ldots\ldots\ldots(2)$$

where $\beta = (p-1)/(p+1)$. Thus

$$e^{-t} \sum_{n=0}^{\infty} (-1)^n L_{2n}(2t) \Rightarrow \frac{p}{p+1} (1 - \beta^2 + \beta^4 - \beta^6 + \ldots), \quad \ldots\ldots(3)$$

$$= \frac{p}{p+1}\left(\frac{1}{1+\beta^2}\right) = \frac{p(p+1)}{2(p^2+1)}, \quad \ldots\ldots(4)$$

provided $|\beta| < 1$ for convergence. Interpreting the r.h.s. of (4), we get

$$\tfrac{1}{2}(\cos t + \sin t) \Rightarrow \frac{p^2}{2(p^2+1)} + \frac{p}{2(p^2+1)}. \quad \ldots\ldots\ldots(5)$$

Hence by (3) and (5)

$$\cos(t - \tfrac{1}{4}\pi) = \sqrt{2}\,e^{-t} \sum_{n=0}^{\infty} (-1)^n L_{2n}(2t), \quad \ldots\ldots\ldots(6)$$

the conditions in § 2·18 being satisfied.

*5·26. Expansion of sin t in terms of Weber's parabolic cylinder function.

From the list

$$e^{\frac{1}{4}t}D_{2n+1}(\sqrt{2t}) \Rightarrow \sqrt{\pi/2}\,(2n+1)!\,(1/p-1)^n/2^n n!\,\sqrt{p}, \quad \ldots\ldots(1)$$

so $\quad \sqrt{2}\,e^{\frac{1}{4}t} \sum_{n=0}^{\infty} (-1)^n \dfrac{2^n}{(2n+1)!} D_{2n+1}(\sqrt{2t})$

$$\Rightarrow \sqrt{\pi/p} \sum_{n=0}^{\infty} (-1)^n (1/p-1)^n/n! \quad \ldots\ldots\ldots(2)$$

$$= \sqrt{\pi/p}\,e^{1-1/p} \subset e\sqrt{\pi t^{\frac{1}{2}}}J_{1/2}(2\sqrt{t}) \quad \ldots\ldots\ldots(3)$$

$$= e \sin 2\sqrt{t}. \quad \ldots\ldots\ldots\ldots(4)$$

Writing t for $2\sqrt{t}$ in (2) and (4), we have the relationship

$$e^{\frac{1}{8}t^2-1} \sum_{n=0}^{\infty} (-1)^n \dfrac{2^{n+\frac{1}{2}}}{(2n+1)!} D_{2n+1}(t/\sqrt{2}) = \sin t, \quad \ldots\ldots(5)$$

the term by term correspondence conditions in § 2·18 being satisfied.

*5·27. A formula of Poisson [ref. 10].

We commence with the long established formula

$$u \coth u = 1 + 2u^2 \sum_{n=1}^{\infty} 1/(u^2 + \pi^2 n^2). \quad \ldots\ldots\ldots(1)$$

Substituting $u = \pi\sqrt{p}$ in (1), we get

$$\pi\sqrt{p} \coth \pi\sqrt{p} = 1 + 2 \sum_{n=1}^{\infty} p/(p+n^2), \quad \ldots\ldots\ldots(2)$$

$$\subset 1 + 2 \sum_{n=1}^{\infty} e^{-n^2 t}, \quad \ldots\ldots\ldots\ldots(3)$$

from the list. The r.h.s. of (3) converges absolutely and uniformly in $t_1 \leqslant t \leqslant t_2$, if $t_1 > 0$.

Also $\quad \coth u = \dfrac{e^u + e^{-u}}{e^u - e^{-u}} = (1 + e^{-2u})(1 - e^{-2u})^{-1}, \quad \ldots\ldots\ldots(4)$

$$= 1 + 2 \sum_{n=1}^{\infty} e^{-2nu}, \quad \ldots\ldots\ldots(5)$$

provided $R(u)>0$. Thus

$$\pi\sqrt{p}\ \coth\ \pi\sqrt{p}=\pi\sqrt{p}\left[1+2\sum_{n=1}^{\infty}e^{-2n\pi\sqrt{p}}\right],\ \ldots\ldots\ldots\ldots(6)$$

$$=\sqrt{\pi/t}\left[1+2\sum_{n=1}^{\infty}e^{-n^2\pi^2/t}\right],\ \ldots\ldots\ldots\ldots(7)$$

the term by term correspondence in (6), (7) being valid, since the conditions in § 2·18 are satisfied.   Hence by Lerch's theorem §§ 1·14, 1·16, (3) and (7) are identical, so

$$1+2\sum_{n=1}^{\infty}e^{-n^2 t}=\sqrt{\pi/t}\left[1+2\sum_{n=1}^{\infty}e^{-\pi^2 n^2/t}\right],\ \ldots\ldots\ldots\ldots(8)^*$$

t being real and >0.   The result holds for $R(t)>0$ also, if t is complex.

* In *Math. and Phys. papers*, Vol. 2, Article 72, Kelvin gave the more general formula

$$\sum_{n=-\infty}^{\infty}e^{-(x+n t^{1/2})^2}=\sqrt{\pi/t}\left[1+2\sum_{n=1}^{\infty}e^{-\pi^2 n^2/t}\cos\frac{2\pi nx}{t^{1/2}}\right].$$

See also Heaviside, *Electrical Papers*, Vol. 1, p. 88.

VI

DERIVATION OF LAPLACE TRANSFORMS OF VARIOUS FUNCTIONS

6·11. In the course of mathematical analysis during the past century or more, many infinite integrals akin to (1) § 1·11 have been evaluated without any regard to their being Laplace transforms. The infinite integral is a standard type employed in integrating various functions.

The L.T. of a function is not always obtained by direct integration of (1) § 1·11. It is frequently expedient to employ various artifices to simplify the analysis. Sometimes an approach via the complex integral part of the Mellin inversion theorem (Appendix IV) may be preferable. When the L.T. of a function has been found, other L.T.S. can be derived from it by aid of the theorems in Chapter II. In what follows, we shall illustrate several methods of obtaining L.T.S. For additional information respecting the derivation of various L.T.S., the references on p. 212 marked * and those in [13] may be consulted.

6·12. Example. Find the L.T. of $t^\nu \log t$, $R(\nu) > -1$.

By § 1·12, we have

$$p \int_0^\infty e^{-pt} t^\nu \, dt = \Gamma(1+\nu)/p^\nu, \quad \dots\dots\dots\dots\dots(1)$$

$R(\nu) > -1$. The integral may be differentiated under the sign with respect to ν, since the resulting integral converges uniformly with respect thereto (see § 12B, Appendix III), so

$$p \int_0^\infty e^{-pt} t^\nu \log t \, dt = \frac{\Gamma'(1+\nu)}{p^\nu} - \frac{\Gamma(1+\nu) \log p}{p^\nu} . \quad \dots\dots(2)\dagger$$

Thus
$$t^\nu \log t \doteqdot \frac{\Gamma(1+\nu)}{p^\nu} [\psi(1+\nu) - \log p], \quad \dots\dots(3)$$

† By § 1·211 the integral is absolutely and uniformly convergent if $R(\nu) > -1$.

where $\psi = \Gamma'/\Gamma = \dfrac{d}{d\nu} \log \Gamma(1+\nu)$. When $\nu = 0$, we have the particular case, with $\gamma = -\psi(1) = 0.5772\ldots$,

$$\log t \rightleftharpoons -\gamma - \log p, \quad\ldots\ldots\ldots\ldots\ldots\ldots(4)$$

or
$$\log(1/t) \rightleftharpoons \gamma + \log p. \quad\ldots\ldots\ldots\ldots\ldots(5)$$

Differentiating (2) with respect to ν, and writing $[\psi(1+\nu) - \log p] = \chi$, gives

$$p \int_0^\infty e^{-pt} t^\nu \log^2 t \, dt$$
$$= \frac{\Gamma(1+\nu)\psi'(1+\nu)}{p^\nu} + \frac{\chi}{p^\nu}[\Gamma'(1+\nu) - \Gamma(1+\nu)\log p], \ldots(6)$$

with $R(\nu) > -1$. Since
$$\psi = \Gamma'/\Gamma \quad \text{and} \quad \psi'(1+\nu) = \zeta(2, \nu),*$$
(6) leads to

$$t^\nu \log^2 t / \Gamma(1+\nu)$$
$$\rightleftharpoons \{[\psi(1+\nu) - \log p]^2 + \zeta(2, \nu)\}/p^\nu = [\chi^2 + \zeta(2, \nu)]/p^\nu. \quad\ldots(7)$$

In particular if $\nu = 0$, (7) gives

$$\log^2 t \rightleftharpoons (\log p + \gamma)^2 + \pi^2/6, \quad\ldots\ldots\ldots\ldots\ldots(8)$$

with $\pi^2/6 = \zeta(2, 0) = \sigma_2$.

Differentiating (6) with respect to ν, we find that

$$t^\nu \log^3 t / \Gamma(1+\nu) \rightleftharpoons [\chi^3 + \zeta(2, \nu)\{2\chi - \log p\} - 2\zeta(3, \nu)]/p^\nu, \quad\ldots(9)$$

$R(\nu) > -1$. In particular with $\nu = 0$,

$$\log^3 t \rightleftharpoons -(\log p + \gamma)^3 - \frac{\pi^2}{6}(2\gamma + 3\log p) - 2\sigma_3, \quad\ldots\ldots(10)$$

where $2\sigma_3 = 2\zeta(3, 0) \simeq \pi^3/12.89718\ldots$.

6·13. Example. Find the L.T. of $\sinh^{-1} t$.

Now $\dfrac{d}{dt} \sinh^{-1} t = 1/\sqrt{1+t^2}$, and from the list

$$p \int_0^\infty e^{-pt} dt / \sqrt{1+t^2} = \frac{\pi p}{2}\left[\mathbf{H}_0(p) - Y_0(p)\right]. \quad\ldots\ldots\ldots(1)$$

Hence by (4) § 2·15

$$\sinh^{-1} t \rightleftharpoons \tfrac{1}{2}\pi[\mathbf{H}_0(p) - Y_0(p)]. \quad\ldots\ldots\ldots\ldots(2)$$

* See [13] for definition of the Zeta function $\zeta(\nu, t)$.

6·14. Example.   Find the L.T.S. of (a) erf $\sqrt{t}$, (b) $J_0(t)$.

(a) $\text{erf } \sqrt{t} = \dfrac{2}{\sqrt{\pi}} \displaystyle\int_0^{\sqrt{t}} e^{-x^2}\, dx.$ ....(1)

By § 12A, Appendix III,

$$\frac{d}{dt}\, \text{erf } \sqrt{t} = \frac{e^{-t}}{\sqrt{\pi t}}.\qquad(2)$$

By (3) § 1·12

$$\frac{1}{\sqrt{\pi t}} \rightleftharpoons \frac{\Gamma(1/2)\sqrt{p}}{\sqrt{\pi}} = \sqrt{p}.\qquad(3)$$

Applying (2) § 2·13 to (3), we obtain

$$e^{-t}/\sqrt{\pi t} \rightleftharpoons \frac{p}{p+1}\sqrt{p+1} = p/\sqrt{p+1}.\qquad(4)$$

Hence by (3) § 2·15

$$\int_0^t e^{-t}\, dt/\sqrt{\pi t} = \text{erf } \sqrt{t} \rightleftharpoons 1/\sqrt{p+1}.\qquad(5)$$

This result may also be obtained by expanding e^{-x^2} in (1), integrating term by term and substituting the L.T.S.

(b) $J_0(t) = \displaystyle\sum_{r=0}^{\infty} (-1)^r (\tfrac{1}{2}t)^{2r}/(r!)^2,$ ....(6)

so $J_0(t) \rightleftharpoons p \displaystyle\int_0^{\infty} e^{-pt}\left[\sum_{r=0}^{\infty} (-1)^r (\tfrac{1}{2}t)^{2r}/(r!)^2 \right] dt.$ ....(7)

It can be shown that the series in (6) is uniformly convergent in the range $t \geqslant 0$ (see § 6, Appendix II), so by § 2·18

$$J_0(t) \rightleftharpoons p \sum_{r=0}^{\infty} (-1)^r \int_0^{\infty} [e^{-pt}(\tfrac{1}{2}t)^{2r}/(r!)^2]\, dt,\qquad(8)$$

$$= \sum_{r=0}^{\infty} (-1)^r \frac{(2r)!}{2^{2r}(r!)^2 p^{2r}} = 1 + \sum_{r=1}^{\infty} (-1)^r \frac{1.3.5 \ldots 2r-1}{2^r r! p^{2r}},\ \ldots(9)$$

$$= p/\sqrt{p^2+1}.\qquad(10)$$

The reader should confirm that conditions (a), (b) § 2·18 are satisfied.   We may also write

$$J_0(it) = I_0(t) = \sum_{r=0}^{\infty} (\tfrac{1}{2}t)^{2r}/(r!)^2 \rightleftharpoons p/\sqrt{p^2-1}.\qquad(11)$$

6·15. Example.   Obtain the L.T.S. of

$$t^{\nu/2}\,\text{ber}_\nu\,\sqrt{t} \text{ and } t^{\nu/2}\,\text{bei}_\nu\,\sqrt{t}, \ \nu \text{ real} > -1.$$

From the list

$$t^{\nu/2}J_\nu(2\sqrt{at}) \Rightarrow a^{\nu/2}e^{-a/p}p^{-\nu}. \quad\text{...............(1)}$$

For a write $\tfrac14 e^{\frac32\pi i}$, and we get

$$t^{\nu/2}J_\nu(i^{3/2}\sqrt{t}) \Rightarrow \frac{i^{3\nu/2}e^{i/4p}}{(2p)^\nu} = \frac{e^{(i/4p)+\frac32\nu\pi i}}{(2p)^\nu}. \quad\text{...............(2)}$$

Thus $t^{\nu/2}(\text{ber}_\nu\,\sqrt{t}+i\,\text{bei}_\nu\,\sqrt{t})$

$$\Rightarrow \frac{1}{(2p)^\nu}\left[\cos\left(\frac{1}{4p}+\frac34\nu\pi\right)+i\sin\left(\frac{1}{4p}+\frac34\nu\pi\right)\right], \quad\text{........(3)}$$

provided the infinite integral of·type (1) § 1·11 corresponding to (2) converges (see 5 § 8·2). This is easily proved to be true if ν real > -1, so separating real and imaginary parts in (3)—assuming p real >0—we obtain

$$t^{\nu/2}\,\text{ber}_\nu\,\sqrt{t} \Rightarrow \frac{1}{(2p)^\nu}\cos\left(\frac{1}{4p}+\frac34\nu\pi\right), \quad\text{...............(4)}$$

and

$$t^{\nu/2}\,\text{bei}_\nu\,\sqrt{t} \Rightarrow \frac{1}{(2p)^\nu}\sin\left(\frac{1}{4p}+\frac34\nu\pi\right). \quad\text{...............(5)}$$

In the particular case when $\nu=0$, we get

$$\text{ber}\,\sqrt{t} \Rightarrow \cos(1/4p), \quad\text{.....................(6)}$$

$$\text{bei}\,\sqrt{t} \Rightarrow \sin(1/4p). \quad\text{.....................(7)}$$

Expanding the r.h.s. of (6) and interpreting term by term, we have •

$$\cos\frac{1}{4p} = \sum_{r=0}^{\infty}(-1)^r\frac{(1/4p)^{2r}}{(2r)!} \subset \sum_{r=0}^{\infty}\frac{(-1)^r(\tfrac12 t)^{2r}}{2^{2r}[(2r)!]^2}. \quad\text{.........(8)}$$

Term by term correspondence in (8) is valid, since the series in p is absolutely convergent, whilst that in t is uniformly convergent in any closed interval $0\leqslant t\leqslant h$ (see § 2·18). Consequently the r.h.s. of (8) is the expansion of ber $\sqrt{t}$. Changing $\sqrt{t}$ to t gives

$$\text{ber}\,t = \sum_{r=0}^{\infty}(-1)^r(\tfrac12 t)^{4r}/[(2r)!]^2. \quad\text{.....................(9)}$$

In like manner it may be established that

$$\text{bei}\,t = \sum_{r=0}^{\infty}(-1)^r(\tfrac12 t)^{4r+2}/[(2r+1)!]^2. \quad\text{.............(10)}$$

*6·16. Example. Given that

$$J_0(a\sqrt{t^2 - b^2}) \rightleftharpoons pe^{-b\sqrt{p^2+a^2}}/\sqrt{p^2 + a^2}, \ a \text{ and } b \text{ real} > 0,$$

obtain the L.T.S. of

$$J_1(a\sqrt{t^2 - b^2})/\sqrt{t^2 - b^2}, \text{ and } \sqrt{t^2 - b^2}\, J_1(a\sqrt{t^2 - b^2}).$$

By definition (2) § 1·31 and the given L.T.

$$p \int_b^\infty e^{-pt} J_0(a\sqrt{t^2 - b^2})\, dt = pe^{-b\sqrt{p^2+a^2}}/\sqrt{p^2 + a^2}. \quad \ldots\ldots(1)$$

Differentiation under the integral sign with respect to b is permissible, since the resulting integral is uniformly convergent (see §§ 12, 13, Appendix III), so we have

$$p \int_b^\infty e^{-pt}\frac{J_1(a\sqrt{t^2 - b^2})}{\sqrt{t^2 - b^2}}\, dt = \frac{p}{ab}(e^{-pb} - e^{-b\sqrt{p^2+a^2}}), \quad \ldots\ldots(2)$$

$$\rightleftharpoons J_1(a\sqrt{t^2 - b^2})/\sqrt{t^2 - b^2}. \quad \ldots\ldots(3)$$

Differentiating (1) under the integral sign with respect to a, leads to the result

$$p \int_b^\infty e^{-pt}\sqrt{t^2 - b^2}\, J_1(a\sqrt{t^2 - b^2})\, dt$$

$$= \frac{pae^{-b\sqrt{p^2+a^2}}}{(p^2 + a^2)}(b + 1/\sqrt{p^2 + a^2}), \quad \ldots\ldots(4)$$

$$\rightleftharpoons \sqrt{t^2 - b^2}\, J_1(a\sqrt{t^2 - b^2}). \quad \ldots\ldots(5)$$

Dividing (2) throughout by p and differentiating the r.h.s. with respect thereto, by (4), (5) § 2·17 we get

$$t\,\frac{J_1(a\sqrt{t^2 - b^2})}{\sqrt{t^2 - b^2}} \rightleftharpoons \frac{p}{a}\left[e^{-pb} - \frac{pe^{-b\sqrt{p^2+a^2}}}{\sqrt{p^2 + a^2}}\right]^{*}. \quad \ldots\ldots(6)$$

The term pe^{-pb} in (2) and (6) is the L.T. of $I(t - b)$, the impulsive function of unit strength which occurs at $t = b$ (see § 2·19). In each case it is neutralised by an equal but opposite term in the expansion of the second function in brackets, since neither of the Bessel functions in (3) and (6) is impulsive. The reader should establish the validity of all three differentiations under the integral sign.

* This may be obtained also by applying (3) § 2·17 to (2) above.

*6·21. Derivation of L.T. from the functional differential equation.

The L.T. can be derived from the differential equation for the function $f(t)$, although this is not *necessarily* the most direct and facile mode of approach. In fact in certain cases it may be cumbersome and protracted, e.g. that illustrated in § 6·23 *et seq.* Nevertheless we shall exemplify the technique, as it may be a useful alternative sometimes. In effect it constitutes a method of solving ordinary linear differential equations with variable coefficients.

The equation to be solved will usually have coefficients expressible as polynomials in terms of the independent variable. Let it take the form

$$g_0(t)\frac{d^n y}{dt^n} + g_1(t)\frac{d^{n-1}y}{dt^{n-1}} + \ \ldots \ + g_n(t)y = 0, \quad \ldots\ldots\ldots\ldots(1)$$

where $g_r(t)$ is a polynomial in t. As in Chapters III and IV, we replace each term in (1) by its transform. With the above equation, having variable coefficients, it is preferable to omit p outside the integral sign in (1) § 1·11, and as in § 2·144 to define the transform thus :

$$y(t) \multimap \Phi(p) = \int_0^\infty e^{-pt}y(t)\,dt. \quad \ldots\ldots\ldots\ldots\ldots(2)$$

Formulae (3) § 2·144 and (1)–(3) § 2·171 are needed to construct from (1) what may be regarded as its transform equation.

*6·22. Example.

Derive the L.T.S. of $\sin t$ and $\cos t$ from the differential equation for the circular functions.

The requisite equation and its formal solution are, respectively,

$$\frac{d^2 y}{dt^2} + y = 0\,; \quad y = A\sin t + B\cos t. \quad \ldots\ldots\ldots\ldots(1)$$

Employing (3) § 2·144 with $y(t) \multimap \Phi(p)$, we have

$$\frac{d^2 y}{dt^2} \multimap p^2\Phi - py(0) - y'(0). \quad \ldots\ldots\ldots\ldots(2)$$

Hence by (1) and (2), the appropriate transform equation is

$$p^2\Phi - py(0) - y'(0) + \Phi = 0. \quad \ldots\ldots\ldots\ldots(3)$$

Now when $y = \sin t$, $y(0) = 0$ and $y'(0) = \cos(0) = 1$, so (3) reduces to

$$(p^2 + 1)\Phi = 1, \quad \text{...(4)}$$

or

$$\Phi = 1/(p^2 + 1), \quad \text{...........................(5)}$$

giving

$$\sin t \rightleftharpoons \phi(p) = p\Phi = p/(p^2 + 1). \quad \text{...................(6)}$$

When $y = \cos t$, $y(0) = 1$, $y'(0) = -\sin(0) = 0$, so (3) yields

$$(p^2 + 1)\Phi = p, \quad \text{.................................(7)}$$

or

$$\cos t \rightleftharpoons p\Phi = p^2/(p^2 + 1). \quad \text{.......................(8)}$$

This analysis should be compared with the procedure in § 2·181. By altering the sign of the first y in (1) above, the L.T.S. of $\sinh t$ and $\cosh t$ can be obtained.

*6·23. Example.   Obtain the L.T. of $I_\nu(t)$, the modified Bessel function of the first kind of order ν, from the appropriate differential equation.

The equation in question is [see 11, (1), p. 113]

$$\frac{d^2 y}{dt^2} + \frac{1}{t}\frac{dy}{dt} - \left(1 + \frac{\nu^2}{t^2}\right)y = 0. \quad \text{.....................(1)}$$

If ν is non-integral, two linearly independent solutions are

$$I_{\pm\nu}(t) = \sum_{r=0}^{\infty} \frac{(\tfrac{1}{2}t)^{\pm\nu+2r}}{r!\,\Gamma(\pm\nu+r+1)}, \quad \text{...................(2)}$$

while an alternative second solution is

$$K_\nu(t) = \frac{\tfrac{1}{2}\pi}{\sin \nu\pi}[I_{-\nu}(t) - I_\nu(t)]. \quad \text{...................(3)}$$

When $\nu = n$, an integer, $I_{-n}(t)$ is not an independent solution, since $I_n = I_{-n}$. The second solution is then $K_n(t)$, the limit of (3) as $\nu \rightarrow n$.

First we write (1) so that all the coefficients involve positive integral powers of t, thereby obtaining

$$t^2 \frac{d^2 y}{dt^2} + t\frac{dy}{dt} - (t^2 + \nu^2)y = 0. \quad \text{.....................(4)}$$

By aid of (3) § 2·144 and (1), (3) § 2·171, the transform terms of (4) are found to be,

$$t^2 \frac{d^2 y}{dt^2} \rightharpoonup p^2\Phi'' + 4p\Phi' + 2\Phi, \quad \text{...................(5)}$$

$$t\frac{dy}{dt} \rightarrow \qquad -p\Phi' - \Phi, \quad \dots\dots\dots(6)$$

$$-t^2y \rightarrow -\Phi'' \qquad , \quad \dots\dots\dots(7)$$

$$-\nu^2 y \rightarrow \qquad -\nu^2\Phi, \quad \dots\dots\dots(8)$$

the dashes signifying differentiation with respect to p. Adding together the r.h. sides of (5)–(8) and equating to zero, we obtain the transform equation

$$(p^2 - 1)\Phi'' + 3p\Phi' + (1 - \nu^2)\Phi = 0. \quad \dots\dots\dots(9)$$

*6·231. Solution of (9) § 6·23.

Let $\Phi = x/u$, where x is a twice differentiable function of p, and $u = \sqrt{p^2 - 1}$, $p > 1$. Then

$$\frac{d\Phi}{dp} = \frac{1}{u}\frac{dx}{dp} - \frac{xp}{u^3} = \frac{1}{u}\left[\frac{dx}{dp} - \frac{xp}{u^2}\right], \quad \dots\dots\dots(1)$$

$$\frac{d^2\Phi}{dp^2} = \frac{1}{u}\frac{d^2x}{dp^2} - \frac{2p}{u^3}\frac{dx}{dp} - \frac{x}{u^3}\left[1 - \frac{3p^2}{u^2}\right]. \quad \dots\dots\dots(2)$$

Substituting from (1) and (2) into (9) § 6·23, leads to the equation

$$(p^2 - 1)\frac{d^2x}{dp^2} + p\frac{dx}{dp} - \nu^2 x = 0. \quad \dots\dots\dots(3)$$

Now write $p = \cosh w$, and

$$\frac{dx}{dp} = \frac{1}{\sinh w} \cdot \frac{dx}{dw}; \quad \frac{d^2x}{dp^2} = \frac{1}{\sinh^2 w}\frac{d^2x}{dw^2} - \frac{\cosh w}{\sinh^3 w} \cdot \frac{dx}{dw}. \quad \dots\dots(4)$$

Substituting from (4) into (3), gives the elementary equation

$$\frac{d^2x}{dw^2} - \nu^2 x = 0, \quad \dots\dots\dots\dots(5)$$

of which the formal solution is

$$x = Ae^{\nu w} + Be^{-\nu w}. \quad \dots\dots\dots\dots(6)$$

Since $w = \cosh^{-1}p = \log(p + \sqrt{p^2 - 1})$, and $\Phi = x/\sqrt{p^2 - 1}$, (6) becomes

$$\Phi = \frac{A(p + \sqrt{p^2 - 1})^\nu}{\sqrt{p^2 - 1}} + \frac{B}{\sqrt{p^2 - 1}(p + \sqrt{p^2 - 1})^\nu}, \quad \dots\dots(7)$$

so $\phi(p) = p\Phi(p)$

$$= \frac{Ap}{\sqrt{p^2 - 1}(p + \sqrt{p^2 - 1})^{-\nu}} + \frac{Bp}{\sqrt{p^2 - 1}(p + \sqrt{p^2 - 1})^\nu}, \quad \dots\dots(8)$$

where A and B are arbitrary constants.

*6·232. Determination of B in (8) § 6·231.

The series for $I_{-\nu}(t)$ at (2) § 6·23, may be derived from that for $I_\nu(t)$ by writing $-\nu$ for ν. Apart from A, B, the two linearly independent solutions in (8), § 6·231, are mutually derivable by the same change. When t is small, if we let $p \to +\infty$ as in 1° § 2·31, the second member of (8), § 6·231, yields

$$B(2p)^{-\nu} \subset B(\tfrac{1}{2}t)^\nu / \Gamma(1+\nu), \quad \dots\dots\dots\dots(1)$$

by (3), § 1·12, provided $R(\nu) > -1$. If $B = 1$, the r.h.s. of (1) is the first term in the expansion of $I_\nu(t)$. Hence we infer that

$$I_\nu(t) = p/\sqrt{p^2-1}\,(p+\sqrt{p^2-1})^\nu, \quad \dots\dots\dots(2)$$

provided $R(\nu) > -1$. The A member in (8) § 6·231 may be treated in a similar way.

Although $I_n(t) = I_{-n}(t)$, n a positive non-zero integer, owing to the above restriction, $I_{-n}(t)$ has no L.T. A similar remark applies to

$$J_{-n}(t) = (-1)^n J_n(t).$$

*6·24. The L.T. of $K_\nu(t)$, a second solution of (1) § 6·23.

By (3) § 6·23 and (2) § 6·232, we have

$$K_\nu(t) = \frac{\pi}{2\sin\nu\pi} \cdot \frac{p}{\sqrt{p^2-1}} \left[\frac{1}{(p+\sqrt{p^2-1})^{-\nu}} - \frac{1}{(p+\sqrt{p^2-1})^\nu}\right]. \quad \dots(1)$$

In (2) § 6·232, $R(\nu) > -1$, so the first bracketed member in (1) holds for $R(\nu) < 1$ only. Consequently the range for the complete formula is $-1 < R(\nu) < 1$. When $\nu = 0$, (1) assumes the indeterminate form $0/0$. Writing (1) as

$$K_\nu(t) = \frac{\pi p}{2\sqrt{p^2-1}} \cdot \frac{1}{\sin\nu\pi} \cdot [(p+\sqrt{p^2-1})^\nu - (p+\sqrt{p^2-1})^{-\nu}], \quad \dots(2)$$

and differentiating the numerator and denominator independently with respect to ν, we obtain

$$\frac{\pi p}{2\sqrt{p^2-1}} \cdot \frac{1}{\pi \cos\nu\pi} \Big[(p+\sqrt{p^2-1})^\nu \log(p+\sqrt{p^2-1})$$
$$+ (p+\sqrt{p^2-1})^{-\nu} \log(p+\sqrt{p^2-1})\Big]. \quad \dots(3)$$

Substituting $\nu = 0$ in (3) gives

$$K_0(t) = \frac{p}{\sqrt{p^2-1}} \log(p+\sqrt{p^2-1}). \quad \dots\dots\dots\dots(4)$$

This result may also be obtained by aid of the relationship

$$-\left[\frac{\partial I_\nu(t)}{\partial \nu}\right]_{\nu=0} = K_0(t).$$

*6·25. The L.T. of $I_\nu(t)$ obtained by the Mellin inversion theorem.

A standard contour integral is [reference 12, (6), p. 105, z and t being interchanged]

$$I_\nu(t) = \frac{(\tfrac{1}{2}t)^\nu}{2\pi i} \int_{-\infty}^{0+} e^{z+t^2/4z}\, dz/z^{\nu+1}, \quad\text{...............}(1)$$

which holds for all values of ν and t. If $R(\nu) > -1$, we may write [12, (7), p. 105]

$$I_\nu(t) = \frac{(\tfrac{1}{2}t)^\nu}{2\pi i} \int_{c-i\infty}^{c+i\infty} e^{z+t^2/4z}\, dz/z^{\nu+1}. \quad\text{...............}(2)$$

The contours for (1) and (2) are illustrated in Fig. 18. Substituting $z + t^2/4z = \zeta t$, or $z = +\tfrac{1}{2}t(\zeta + \sqrt{\zeta^2 - 1})$, we have

$$dz/z = d\zeta/\sqrt{\zeta^2 - 1}$$

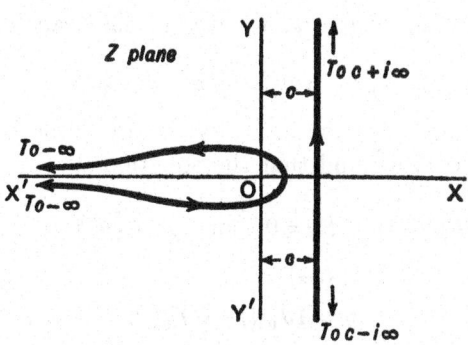

FIG. 18.—The contours $c \pm i\infty$, and $-\infty$, $0+$, $-\infty$, which are, respectively, Br_1, Br_2 of reference 12.

[ref. 12, pp. 65, 77].   Consequently (2) takes the form

$$I_\nu(t) = \frac{1}{2\pi i} \int_{c-i\infty}^{c+i\infty} e^{\zeta t}\, d\zeta/\sqrt{\zeta^2 - 1}\,(\zeta + \sqrt{\zeta^2 - 1})^\nu, \quad\text{...........}(3)$$

so by the Mellin theorem, Appendix IV, with $p>1$,

$$I_\nu(t) \Rightarrow \phi(p) = p/\sqrt{p^2 - 1}\,(p + \sqrt{p^2 - 1})^\nu. \quad\text{...............}(4)$$

This analysis is briefer and neater than that given in §§ 6·23–6·232 ; a not uncommon occurrence where complex integration is concerned.

As exercises, the reader should reproduce some (or all!) of the L.T.S. in the list. Occasionally it will be expedient to obtain results by aid of the Mellin theorem. In the expansion of a L.T. obtained from an integral like (3), in all terms of the form p^ν, $R(\nu) < 1$. Otherwise the corresponding contour integrals would diverge, e.g. there is no integral of type (3) corresponding to $\phi(p) = p$. The Mellin theorem is not applicable when $R(\nu) \geqslant 1$ [see ref. 12, p. 352 and Appendix 7].

*6·31. Solution of differential equation with variable coefficients by inversion. The procedure in §§ 6·22–6·232 provides a means of solving linear equations by direct application of the L.T. We shall now show how to solve an equation by an inversion process using the L.T. Let the equation be

$$(u^2 + 1)\frac{d^2w}{du^2} + 3u\frac{dw}{du} + w = 0. \quad \dots\dots\dots(1)$$

Writing p for u, Φ for w, we obtain the transform *type* of equation

$$(p^2 + 1)\Phi'' + 3p\Phi' + \Phi = 0 \dots\dots\dots(2)$$

Next we get the equation in t, of which (2) is the L.T. version. By (5)–(8) § 6·23 we find that the equation is

$$t^2y'' + ty' + t^2y = 0, \quad \text{or} \quad y'' + \frac{1}{t}y' + y = 0. \quad \dots\dots(3)$$

Then by reference [11, (34), p. 7]

$$y = AJ_0(t) + BY_0(t), \quad \dots\dots\dots(4)$$

where A, B are arbitrary constants. From the list, the transform version of (4) is, absorbing the factor $-2/\pi$ into B,

$$\Phi = \phi/p = \frac{A}{\sqrt{p^2+1}} + \frac{B}{\sqrt{p^2+1}} \log_e(p + \sqrt{p^2+1}). \quad \dots\dots(5)$$

Hence restoring the original variables, the solution of (1) is

$$w = \frac{A}{\sqrt{u^2+1}} + \frac{B}{\sqrt{u^2+1}} \log_e(u + \sqrt{u^2+1}). \quad \dots\dots(6)$$

6·41. The type of function represented by ϕ (p). Many of the functions in the list and in reference 13 are monotonic for p real. The L.T. of a monotonic function $f(t)$ must itself be monotonic. The L.T. of an alternating function may, however, be monotonic, e.g. $\sin t \rightleftharpoons p/(p^2+1)$, $J_0(t) \rightleftharpoons p/\sqrt{p^2+1}$. This occurs, since in both cases $e^{-pt}f(t)$ is an alternating function such that any positive area between it and the t axis is greater than the succeeding negative area. Thus $\overset{\infty}{\Sigma}$ positive areas $> \overset{\infty}{\Sigma}$ negative areas. The same holds for $f(t) = e^{at} \sin t$, $e^{at}J_0(t)$, where a is real > 0, $p > a$. If $\phi(p)$ is an alternating function, it follows that $f(t)$ alternates also, e.g.

$$t^{-1/2} \sin \sqrt{2t} \sinh \sqrt{2t} \rightleftharpoons \sqrt{\pi p} \sin 1/p \ ; \ \text{ber} \ 2\sqrt{t} \rightleftharpoons \cos 1/p.$$

Since $f(at) \rightleftharpoons \phi(p/a)$, if the latter has an exponential factor of the type $e^{-a\sqrt{p}}$, $e^{-a/p}$, etc., and integral (1) § 1·11 converges when a is imaginary or complex, then $\phi(p/a)$ will be an alternating function. For instance

$$t^{-1/2} e^{-a^2/4t} \rightleftharpoons \sqrt{\pi p} \, e^{-a\sqrt{p}} \quad \dots\dots\dots\dots\dots\dots(1)$$

Now (1) § 1·11 converges when $f(t) = t^{-1/2}e^{-a^2/4t}$ and $a^2 = i$, so with $a = +\sqrt{i} = \dfrac{1}{\sqrt{2}}(1+i)$, p real > 0, (1) gives

$$t^{-1/2} e^{-i/4t} \rightleftharpoons \sqrt{\pi p} \, e^{-(1+i)\sqrt{p/2}}. \quad \dots\dots\dots\dots\dots(2)$$

Separating real and imaginary parts, we get

$$t^{-1/2} \cos (1/4t) \rightleftharpoons \sqrt{\pi p} \, e^{-\sqrt{p/2}} \cos \sqrt{p/2}, \quad \dots\dots\dots\dots(3)$$

$$t^{-1/2} \sin (1/4t) \rightleftharpoons \sqrt{\pi p} \, e^{-\sqrt{p/2}} \sin \sqrt{p/2}. \quad \dots\dots\dots\dots(4)$$

Another example is given in § 6·15. In (3), (4) the interval between consecutive zeros of $f(t)$ and of $\phi(p)$ increases with increase in t and p, respectively. Neither function is periodic, since it does not repeat itself exactly at a constant interval.

* In terms of a complex variable, $\phi(p)$ is an analytic (regular or holomorphic) function of p in that part of the p-plane on the right of the contour $c \pm i\infty$. See Figs. 4a, 18, and Appendix IV. Singularities are treated in [12].

As a matter of interest, the reader should plot the functions
on both sides of (3), (4).

Functions involving $e^{a\sqrt{t}}$, $e^{a/t}$ may yield an exponential factor
when the Laplace integral is written in a form suitable for
evaluation.   Thus with $t = v^2/\sqrt{p}$,

$$p \int_0^\infty e^{-pt-1/t}\, dt/t^{1/2} = 2e^{-2\sqrt{p}}p^{3/4} \int_0^\infty e^{-\sqrt{p}(v-1/v)^2}\, dv, \quad \ldots\ldots(5)$$

$$= \sqrt{\pi p}\, e^{-2\sqrt{p}}, \qquad \ldots\ldots\ldots\ldots\ldots\ldots(6)$$

which gives the factor $e^{-2\sqrt{p}}$.

VII

LAPLACE TRANSFORM FOR A FINITE INTERVAL : IMPULSES

7·11. Example. Find the L.T. of sin ωt for the interval $t = m\pi/\omega$ to $n\pi/\omega$, $n > m$, both being positive integers.

In this case the L.T. is found without the aid of special formulae deduced in § 7·12. Referring to Fig. 19, the function

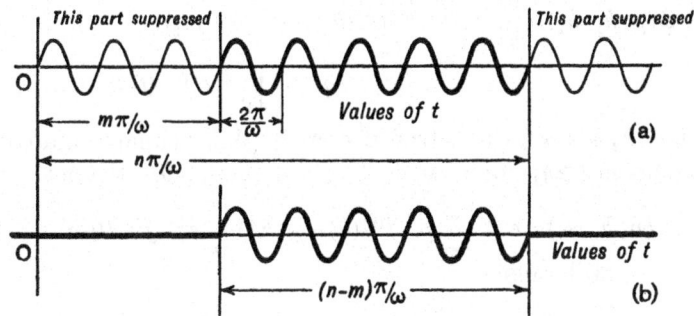

FIG. 19.—(b).—The bounded, continuous function

$$f(t) = 0 \quad \left. \begin{array}{l} 0 \leqslant t \leqslant m\pi/\omega, \ \dfrac{n\pi}{\omega} \leqslant t < \infty \\[2mm] = \sin \omega t \end{array} \right\} \ m\pi/\omega \leqslant t \leqslant n\pi/\omega, \ n > m.$$

must be suppressed from $t = 0$ to $m\pi/\omega$, and from $t = n\pi/\omega$ to $+ \infty$. Moving the origin to $t = m\pi/\omega$, the L.T. of sin ωt from this point to $+ \infty$ is, by § 2·12 and the list,

$$\phi(p \ ; \ m\pi/\omega, \ \infty \ ) = \pm e^{-m p\pi/\omega} \omega p/(p^2 + \omega^2), \quad \ldots \ldots \ldots \ldots (1)$$

according as m is even or odd. Likewise the L.T. from $n\pi/\omega$ to $+ \infty$ is

$$\phi(p \ ; \ n\pi/\omega, \ \infty \ ) = \pm e^{-n p\pi/\omega} \omega p/(p^2 + \omega^2), \quad \ldots \ldots \ldots \ldots (2)$$

according as n is even or odd. The L.T. from $t = m\pi/\omega$ to $n\pi/\omega$ is the difference between (1) and (2), namely,

$$\phi(p \ ; \ m\pi/\omega, \ n\pi/\omega) = \pm \omega p \, (e^{-m p\pi/\omega} - e^{-n p\pi/\omega})/(p^2 + \omega^2), \quad \ldots (3)$$

according as m, n are both even or both odd. If m is even but n is odd, we have

$$\phi(p \ ; \ m\pi/\omega, \ n\pi/\omega) = \omega p \, (e^{-m p\pi/\omega} + e^{-n p\pi/\omega})/(p^2 + \omega^2). \quad \ldots \ldots (4)$$

7·12. General formulae. If $f(t)$ is continuous in the finite interval $h_1 \leqslant t \leqslant h_2, {}^* h_1 \geqslant 0$, but may be finitely discontinuous at h_1, h_2, e.g. Fig. 22, we define the L.T. for the interval to be

$$\phi(p; h_1, h_2) = p \int_{h_1}^{h_2} e^{-pt} f(t)\, dt. \qquad\qquad (1)$$

Integrating (1) by parts once, and inserting the limits, leads to

$$\phi(p; h_1, h_2) = e^{-ph_1} f(h_1) - e^{-ph_2} f(h_2) + \int_{h_1}^{h_2} e^{-pt} f'(t)\, dt, \quad\dots (2)$$

provided $f'(t)$ is continuous in $h_1 \leqslant t \leqslant h_2$. Repeating the process of partial integration, if $f(t)$ and its first n derivatives are continuous† in $h_1 \leqslant t \leqslant h_2$, $h_1 \geqslant 0$, we find that

$$\phi(p; h_1, h_2) = \sum_{r=0}^{n-1} p^{-r} [e^{-ph_1} f^{(r)}(h_1) - e^{-ph_2} f^{(r)}(h_2)]$$

$$+ p^{-n+1} \int_{h_1}^{h_2} e^{-pt} f^{(n)}(t)\, dt. \qquad\dots (3)$$

When $h_2 \to +\infty$, (3) is true if $e^{-pt} f^{(r)}(t)$ is bounded and continuous in $t \geqslant h_1$, for $r = 0, 1, 2, \dots n$, $p > p_0 > 0$. We get

$$\phi(p; h, \infty) = e^{-ph} \sum_{r=0}^{n-1} p^{-r} f^{(r)}(h) + p^{-n+1} \int_{h}^{\infty} e^{-pt} f^{(n)}(t)\, dt. \quad\dots (4)$$

If $h = 0$, (4) becomes

$$\phi(p) = p \int_{0}^{\infty} e^{-pt} f(t)\, dt = \sum_{r=0}^{n-1} p^{-r} f^{(r)}(0) + p^{-n+1} \int_{0}^{\infty} e^{-pt} f^{(n)}(t)\, dt. \quad\dots (5)$$

This is valid for $n = 1$, so $\phi(p) \to f(0)$ if the integral vanishes as $p \to +\infty$, i.e. $\lim\limits_{p \to +\infty} \phi(p) = f(0)$. Taking $h_1 = 0$ and $h_2 = h > 0$, in (3), gives

$$\phi(p; 0, h) = \sum_{r=0}^{n-1} p^{-r} [f^{(r)}(0) - e^{-ph} f^{(r)}(h)]$$

$$+ p^{-n+1} \int_{0}^{h} e^{-pt} f^{(n)}(t)\, dt. \qquad\dots (6)$$

---

* See, for instance, Fig. 20, where $f(t) = t^3$ is continuous in $0 \leqslant t \leqslant h$. If the function defined by $\begin{array}{l} F(t) = t^3 \\ = 0 \end{array} \left.\begin{array}{l} 0 < t < h \\ t < 0, t > h \end{array}\right\}$ is repeated indefinitely with period h, finite discontinuities occur at $t = nh$, n being a positive integer. At any point t_1 within the interval $t = 0$ to h, continuity implies that $F(t_1 - 0)$, $F(t_1)$, $F(t_1 + 0)$ exist and are equal. In the present instance $f(t)$, but not $F(t)$, must be continuous at the end points $t = 0, h$; so $f(0 - 0)$, $f(0)$, $f(0 + 0)$ must exist and be equal, likewise $f(h - 0)$, $f(h)$, $f(h + 0)$.

† Continuity of $f^{(n)}(t)$ implies that of $f^{(r)}(t)$, $r = 0, 1, 2, \dots n - 1$.

Since $\phi(p; 0, h) = \phi(p; 0, \infty) - \phi(p; h, \infty)$, from (4), (5) we obtain

$$\phi(p; 0, h) = \phi(p) - e^{-ph} \sum_{r=0}^{n-1} p^{-r} f^{(r)}(h) - p^{-n+1} \int_h^{\infty} e^{-pt} f^{(n)}(t) \, dt. \quad \ldots(7)$$

Writing $f^{(n)}(t) \rightleftharpoons \phi_n(p)$, (4) may be expressed in the form

$$\phi_n(p; h, \infty) = p^n \phi(p; h, \infty) - e^{-ph} \sum_{r=0}^{n-1} p^{n-r} f^{(r)}(h). \quad \ldots\ldots(8)$$

From (6)

$$\phi_n(p; 0, h) = p^n \phi(p; 0, h) - \sum_{r=0}^{n-1} p^{n-r} [f^{(r)}(0) - e^{-ph} f^{(r)}(h)]. \quad (9)$$

Adding (8), (9) we reproduce (8) § 2·14, namely,

$$\phi_n(p) = p^n \phi(p) - \sum_{r=0}^{n-1} p^{n-r} f^{(r)}(0). \quad \ldots\ldots\ldots\ldots(10)$$

7·13. Example. Obtain the L.T. of $F(t) = t^3 \left. \right\} 0 \leqslant t < h$, h finite, $= 0 \left. \int \right. t \leqslant 0$, $t > h$.

Now t^3 and its first derivative $3t^2$ are continuous in the interval $0 \leqslant t \leqslant h$, so by (7) § 7·12 and (3) § 1·12, we have

$$\phi(p; 0, h) = \frac{3!}{p^3} - e^{-ph} \left(h^3 + \frac{3h^2}{p} + \frac{6h}{p^2} \right) - \frac{3!}{p^2} \int_h^{\infty} e^{-pt} \, dt. \quad \ldots(1)$$

Thus $\quad F(t) \rightleftharpoons \dfrac{3!}{p^3}(1 - e^{-ph}) - e^{-ph} \left(h^3 + \dfrac{3h^2}{p} + \dfrac{6h}{p^2} \right), \quad \ldots\ldots(2)$

and when $h \to +\infty$, this degenerates to

$$F(t) = f(t) \rightleftharpoons \phi(p) = 3!/p^3. \quad \ldots\ldots\ldots\ldots(3)$$

The function $F(t) = t^3 \left. \right\} 0 \leqslant t < h, \left. \atop 0 \int t \leqslant 0, t > h \right\}$, which may be considered to represent an impulsive force* whose magnitude and duration are finite, is depicted in Fig. 20. It is finitely discontinuous at $t = h$. The strength of the impulse, *as distinguished from the impulsive force*, is defined to be

$$\int_0^h F(t) \, dt, \quad \ldots\ldots\ldots\ldots\ldots(4)$$

i.e. the shaded area in Fig. 20.

* That is to say in a practical application where $F(t)$ is the functional form of a force applied to a system.

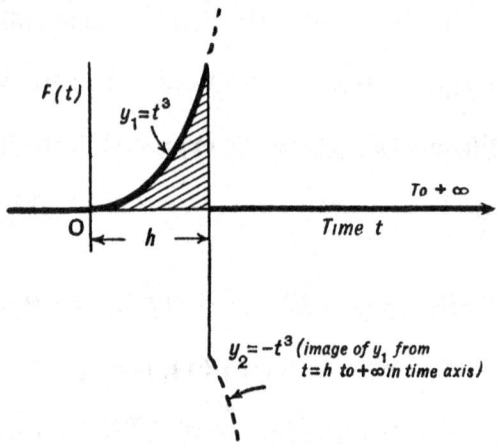

FIG. 20.—The bounded function

$$F(t) = t^3 \left.\begin{matrix} \\ \end{matrix}\right\} \begin{matrix} 0 \leqslant t < h \\ \end{matrix}$$
$$\qquad = 0 \left.\begin{matrix} \\ \end{matrix}\right\} \begin{matrix} t < 0, \; t > h \end{matrix}$$

It is continuous in $t \geqslant 0$, except at the finite discontinuity $t = h$. In $t < 0$ and $t > h$ it is zero.

$F(t)$ may be regarded as the sum of two functions, namely, $y_1 = t^3$, $0 \leqslant t < \infty$, and $y_2 = -t^3$, $h < t < \infty$. Moreover, the third and fourth members of (1) represent the L.T. of y_2, i.e.

$$y_2 = \begin{matrix} -t^3 \\ 0 \end{matrix} \left.\begin{matrix} \\ \end{matrix}\right\} \begin{matrix} h < t < \infty \\ t < h \end{matrix} \left.\begin{matrix} \\ \end{matrix}\right\} \Rightarrow -e^{-ph}\left(h^3 + \frac{3h^2}{p} + \frac{6h}{p^2} + \frac{6}{p^3}\right). \quad \ldots\ldots(5)$$

7·14.   Periodic impulses of finite duration : Fourier expansions.

When an impulse of the type contemplated in § 7·13 is repeated indefinitely (see Fig. 21a), the corresponding function of t can be expressed as a Fourier expansion, provided certain conditions are satisfied [3, 4, 16]. Throughout the interval $t = 0$ to h, the function (a) must be single valued except at its discontinuities of the type in 1° § 1·15;* (b) the discontinuities, maxima, and minima must be limited in number. Let the L.T. of the zeroth period from $t = 0$ to h be $\phi(0, h)$—omitting p for

---

* See 1° § 1·15.   These are in effect the conditions specified in 1829 by Dirichlet (a pupil of Fourier) for the expansion of a function in a Fourier series.   Less stringent conditions are sufficient [4], but those given above are suitable herein.

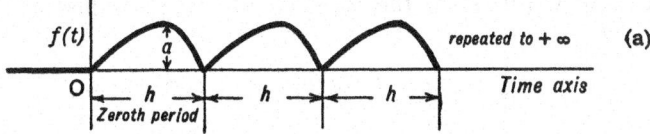

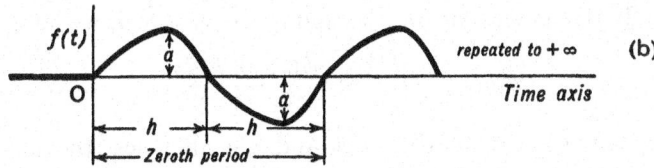

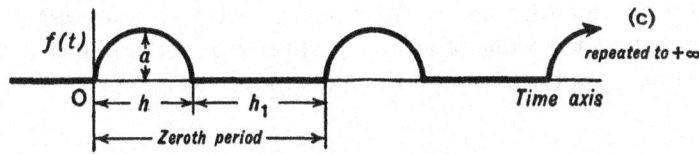

FIG. 21 (a, c).—Illustrating a positive periodic function.
(b).—Illustrating an alternating periodic function.

These functions are bounded, and continuous in $t \geqslant 0$.

brevity—then that of the nth period is $e^{-nph}\phi(0, h)$. Assuming the preceding conditions to be satisfied, by (1) § 7·12 we have

$$\phi(0, \infty) = \phi(0, h) \sum_{n=0}^{\infty} e^{-nph} = \left[p \int_{0}^{h} e^{-pt} f(t)\, dt \right] \sum_{n=0}^{\infty} e^{-nph}. \quad \ldots(1)$$

For the series to converge we must have $|e^{-nph}| < 1$, and since we postulate $R(p) > 0$, this condition is satisfied. Thus

$$\phi(0, \infty) = \phi(0, h)/(1 - e^{-ph}) = \left[p \int_{0}^{h} e^{-pt} f(t)\, dt \right] \Big/ (1 - e^{-ph}). \quad (2)$$

Then the Fourier expansion of the repeated (periodic) function is the inverse of (2) in terms of t for the range $t > 0$. Since the period of the function is h, so also is that of the series. Hence the *series* will have the same sets of values for the intervals $(-h, 0)$, $(-2h, -h)$, ... as for the interval $(0, h)$, and represents the function in $t < 0$, except at discontinuities.

We may write (2) in the form of an integral equation for $y(t)$, thus

$$p \int_0^\infty e^{-pt} y(t)\,dt = \phi(0, h)/(1 - e^{-ph}), \quad \ldots\ldots\ldots\ldots(3)$$

$y(t)$ being the Fourier expansion. The simplest method of solving (3) is to use the Mellin inversion theorem (see Appendix IV). If the conditions for its validity are satisfied, we have

$$y(t) = \frac{1}{2\pi i} \int_{c-i\infty}^{c+i\infty} e^{zt} \frac{\phi(z\,;\ 0, h)\,dz}{z(1 - e^{-zh})} . \quad \ldots\ldots\ldots\ldots(4)$$

The singularities of the integrand are all poles, the factor $1/(1 - e^{-zh})$ contributing an infinite number of them.* In evaluating an integral of this type, where the poles extend to infinity on either side of the imaginary axis, it is necessary to show that the value of the integral taken round a semicircle of radius r_m on the left of $c \pm i\infty$ tends to zero as $r_m \to \infty$.

7·141. Alternate impulses reversed. The zeroth period now extends from $t = 0$ to $2h$ (see Fig. 21b). The L.T. from

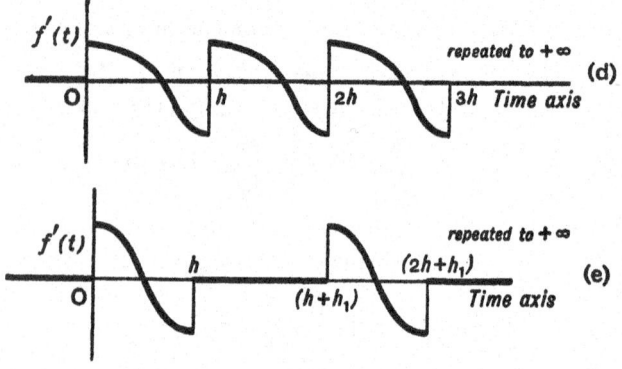

Fig. 21 (d), (e).—Periodic piecewise continuous functions representing, respectively, the first derivatives of Figs. 21 (a), (c). They are bounded in $t \geqslant 0$.

$t = 0$ to h is $\phi(0, h)$, that from $t = h$ to $2h$ is $-\phi(0, h)e^{-ph}$, so

$$\phi(0, 2h) = \phi(0, h)[1 - e^{-ph}]. \quad \ldots\ldots\ldots\ldots(1)$$

* In a strict sense, there is a finite number of poles and a limit point at infinity. See below, (7) § 7·16.

Therefore by (2) § 7·14

$$\phi(0, \infty) = \phi(0, h)[1 - e^{-ph}]/(1 - e^{-2ph}), \quad\dots\dots\dots(2)$$
$$= \phi(0, h)/(1 + e^{-ph}). \quad\dots\dots\dots(3)$$

7·142. Impulse followed by quiescent interval. If the impulse from $t = 0$ to h is followed by a quiescent interval h_1 (see Fig. 21c), the L.T. of the zeroth period is $\phi(0, h + h_1)$, and that of the periodic impulse becomes by (2) § 7·14,

$$\phi(0, \infty) = \phi(0, h + h_1)/[1 - e^{-p(h + h_1)}]. \quad\dots\dots\dots(1)$$

When alternate impulses of this type are reversed, we have

$$\phi(0, \infty) = \phi(0, h + h_1)/[1 + e^{-p(h + h_1)}]. \quad\dots\dots\dots(2)$$

the period being $2(h + h_1)$. In the above cases and in § 7·141, the Fourier expansion is obtained by evaluating an integral of the type (4) § 7·14.

7·15. Example. Find the Fourier expansion of a semi-infinite sequence of dots in the Morse telegraph code (see Fig. 22). This may be regarded as the 'Morse dot function'.

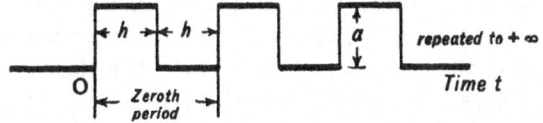

Fig. 22.—The Morse dot function. It is positive, and periodic piece-wise continuous in $t \geqslant 0$. Finite discontinuities occur at $t = nh$, n a positive integer.

In the interval $t = (0, h)$ the function is $aH(t)$, while in $t = (0, 2h)$ it is $a[H(t) - H(t - h)]$, being zero in $h < t < 2h$.

Now
$$aH(t) \rightleftharpoons a$$
and
$$aH(t - h) \rightleftharpoons ae^{-ph}$$
$$\bigg\}, \quad\dots\dots\dots\dots\dots(1)$$

so for the zeroth period $t = (0, 2h)$, we have

$$a[H(t) - H(t - h)] \rightleftharpoons a(1 - e^{-ph}) = \phi(0, 2h). \dots\dots\dots(2)$$

Accordingly by (1) § 7·142

$$\phi(0, \infty) = a/(1 + e^{-ph}). \quad\dots\dots\dots\dots(3)$$

Applying (4) § 7·14 to (3) gives

$$f(t) = \frac{a}{2\pi i} \int_{c - i\infty}^{c + i\infty} e^{zt} \, dz/z(1 + e^{-zh}). \quad\dots\dots\dots(4)$$

It may be shown that integral (4) taken round a large semi-circle of radius r_m on the left of $c \pm i\infty$ (Fig. 18), which meets this contour *between* the mth and $(m+1)$th poles, tends to zero as $r_m \to \infty$ (see remarks below (7) § 7·16). Then the Fourier expansion of the Morse dot function is obtained by summing the residues at all the poles of the integrand in (4). The poles occur at $z = 0$, $zh = (2n+1)\pi i$, $n = -\infty$ to ∞.* Hence by the theorem of residues [12, p. 45], we obtain

$$f(t) = \frac{a}{2} + a \sum_{n=-\infty}^{\infty} \left[\frac{e^{zt}}{z \dfrac{d}{dz}(1 + e^{-zh})} \right]_{zh=(2n+1)\pi i} \quad \ldots\ldots\ldots\ldots(5)$$

$$= \frac{1}{2}a - a \sum_{n=-\infty}^{\infty} \left[\frac{e^{zt}}{zhe^{-zh}} \right]_{zh=(2n+1)\pi i} \quad \ldots\ldots\ldots\ldots(6)$$

$$= a \left[\frac{1}{2} + \frac{2}{\pi} \sum_{n=0}^{\infty} \frac{\sin (2n+1)\pi t/h}{2n+1} \right]. \quad \ldots\ldots\ldots\ldots(7)$$

The term $\frac{1}{2}a$ represents the mean value of the dot function. By applying Dirichlet's test [3, pp. 59, 60, 127 ; 4, p. 152] it can be shown that the infinite series in (7) is *uniformly* convergent in the intervals $0 < t < h$; $h < t < 2h$; etc., where for convenience we have used an open interval. In these intervals the series term represents a continuous function. For the first interval this function is $+\frac{1}{2}$, and for the second it is $-\frac{1}{2}$. Thus

$$\frac{2}{\pi} \sum_{n=0}^{\infty} \frac{\sin (2n+1)\pi t/h}{2n+1} = +\frac{1}{2}, \quad 0 < t < h, \quad \ldots\ldots\ldots\ldots(8)$$

and

$$\frac{2}{\pi} \sum_{n=0}^{\infty} \frac{\sin (2n+1)\pi t/h}{2n+1} = -\frac{1}{2}, \quad h < t < 2h. \quad \ldots\ldots\ldots(9)$$

The series is ordinarily convergent at the points $t = 0, h, 2h, \ldots$ where it has the value zero. The dot function is discontinuous at these points, and is not represented by (7) thereat.

7·16. Example. An electrical generator of resistance R is connected in series with a capacitance C, as shown schematically in Fig. 23. The p.d. developed by the generator is given by $E = E_0 \, | \sin \omega t \, |$. If C is uncharged and the switch closed at $t = 0$, $E = 0$, what is the current at any subsequent time?

---

* The so-called point at infinity ($|z|$ infinite), is a limit point of the poles. [16].

The differential equation for the circuit is

$$RI + \frac{1}{C}\int_0^t I\,dt = E_0 \,|\sin \omega t\,|, \qquad \dots\dots\dots(1)$$

I being the current, which is a function of t. The transform equation corresponding to (1) is by § 3·18

$$\phi(R + 1/pC) = E_0\phi_1 \subset E_0\,|\sin \omega t\,|, \qquad \dots\dots(2)$$

since the system is quiescent initially. The generator wave form is the 'rectified' version* of $\sin \omega t$, i.e. the first half period is repeated indefinitely. The L.T. of $\sin \omega t$ is $\omega p/(p^2 + \omega^2)$, whilst that of the same function commencing at $t = \pi/\omega$ is, by § 2·12,

$$e^{-\pi p/\omega}\omega p/(p^2 + \omega^2).$$

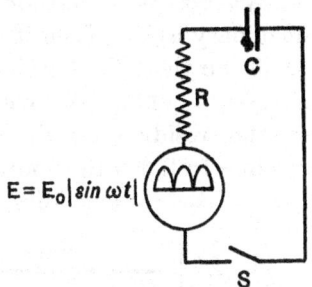

$$E = E_0|\sin \omega t|$$

Fig. 23.—Schematic diagram of simple resistance-capacitance circuit.

The sum of these functions is the L.T. of the first semi-period of $\sin \omega t$. Thus

$$\phi(0,\,\pi/\omega) = \frac{\omega p}{(p^2 + \omega^2)}(1 + e^{-\pi p/\omega}), \qquad \dots\dots\dots(3)$$

and by (2) § 7·14

$$\phi_1 = \frac{\omega p}{(p^2 + \omega^2)}\left(\frac{1 + e^{-\pi p/\omega}}{1 - e^{-\pi p/\omega}}\right). \qquad \dots\dots\dots(4)$$

Substituting from (4) into (2), we obtain

$$\phi = \frac{E_0\omega C p^2}{(p^2 + \omega^2)(pCR + 1)}\left(\frac{1 + e^{-\pi p/\omega}}{1 - e^{-\pi p/\omega}}\right). \qquad \dots\dots(5)$$

Since (5) satisfies the requisite conditions for applying the Mellin theorem (Appendix IV), we have

$$I = \frac{(E_0/R)\omega}{2\pi i}\int_{c-i\infty}^{c+i\infty}\frac{e^{zt}(1 + e^{-\pi z/\omega})z\,dz}{(z^2 + \omega^2)(z + a)(1 - e^{-\pi z/\omega})}, \qquad \dots\dots(6)$$

where $a = 1/CR$. The singularities of the integrand are poles at $z = -a$, $z = 2\omega n i$, $n = -\infty$ to ∞, excluding zero in virtue of z in the numerator. Owing to the factor $(1 + e^{-\pi z/\omega})$, there are no

* In an electrical sense, the p.d. being unidirectional but pulsating.

poles due to $(z^2 + \omega^2)$. Omitting $E_0\omega/R$, the contribution from the pole $z = -a$, is

$$\frac{-a\,e^{-at}}{(\omega^2 + a^2)}\left(\frac{1 + e^{a\pi/\omega}}{1 - e^{a\pi/\omega}}\right). \quad \dots\dots\dots\dots(7)$$

In order to evaluate (6) along the path $c \pm i\infty$, when there is an infinite number of simple poles, we take a large semicircle of radius r_m on the left of $c \pm i\infty$, which meets it in two points *between* the mth and $(m+1)$th poles on either side of the imaginary axis. Then if we imagine m and r_m to $\to\infty$, it can be shown that the contribution to the integral from the semi-circle $\to 0$. Thus the contribution from $c \pm i\infty$ is the sum of the residues at the ·poles. Hence by the theorem of residues [ref. 12] the contribution from the poles at $z = 2\omega ni$, $n = -\infty$ to ∞,* excluding $n = 0$, is

$$\sum_{n=-\infty}^{\infty}{}' \left[\frac{e^{zt}z(1 + e^{-\pi z/\omega})}{(z^2 + \omega^2)(z + a)\dfrac{d}{dz}(1 - e^{-\pi z/\omega})}\right]_{z=2\omega ni} \quad \dots\dots\dots(8)\dagger$$

$$= \frac{\omega}{\pi}\sum_{n=-\infty}^{\infty}{}' \left[\frac{e^{zt}z(1 + e^{-\pi z/\omega})}{(z^2 + \omega^2)(z + a)e^{-\pi z/\omega}}\right]_{z=2\omega ni}. \quad \dots\dots\dots\dots(9)$$

$$= \frac{\omega}{\pi}\sum_{n=1}^{\infty} \left[\frac{e^{2\omega nti}\,4\omega ni}{\omega^2(1 - 4n^2)(a + 2\omega ni)} - \frac{e^{-2\omega nti}\,4\omega ni}{\omega^2(1 - 4n^2)(a - 2\omega ni)}\right], \quad \dots(10)$$

$$= \frac{8}{\pi}\sum_{n=1}^{\infty} \left(\frac{n}{4n^2 - 1}\right)\left[\frac{a\sin 2\omega nt - 2\omega n\cos 2\omega nt}{a^2 + 4\omega^2 n^2}\right], \quad \dots\dots\dots(11)$$

$$= \frac{8CR}{\pi}\sum_{n=1}^{\infty} \left(\frac{n}{4n^2 - 1}\right)\frac{\sin(2\omega nt - \tan^{-1} 2\omega nCR)}{\sqrt{(1 + 4\omega^2 n^2 C^2 R^2)}}. \quad \dots\dots\dots(12)$$

Adding (7), (12) and inserting the external factor $\omega E_0/R$ from (6), the circuital current is given by

$$I = E_0\omega C\left[\frac{8}{\pi}\sum_{n=1}^{\infty}\left(\frac{n}{4n^2 - 1}\right)\frac{\sin(2\omega nt - \tan^{-1} 2\omega nCR)}{\sqrt{(1 + 4\omega^2 n^2 C^2 R^2)}} - \right.$$
$$\left. - \frac{e^{-t/CR}(1 + e^{\pi/\omega CR})}{(1 + \omega^2 C^2 R^2)(1 - e^{\pi/\omega CR})}\right]. \quad \dots(13)$$

The modulus of each term of the series in (12), (13) $\leqslant K/(4n^2 - 1)$,

---

* See footnote to § 7·15.

† The dash on Σ implies omission of $n = 0$.

where K is a constant independent of t. Now $K \sum\limits_{n=1}^{\infty} 1/(4n^2 - 1)$ is convergent, since $\sum\limits_{n=1}^{\infty} 1/n^2$ is convergent. Hence by the M test in 2° § 6, Appendix II, the series is absolutely and uniformly convergent for all t, and represents a function continuous in $t \geqslant 0$. This series is the Fourier expansion of the current wave *after* the transient has died away (in a practical sense). The transient is represented by the second term (with exponentials) in (13). Due to the presence of C, the current is ultimately alternating in character, there being no unidirectional component.

7·17. Repeated impulses of infinite amplitude.

If an impulse of the type considered in § 2·19, where $AI(t) \Rightarrow Ap$, is repeated at interval h (see Fig. 24a), then by (2) § 7·14 the L.T. is

$$\phi = Ap/(1 - e^{-ph}). \quad\dots\dots\dots\dots\dots\dots\dots(1)$$

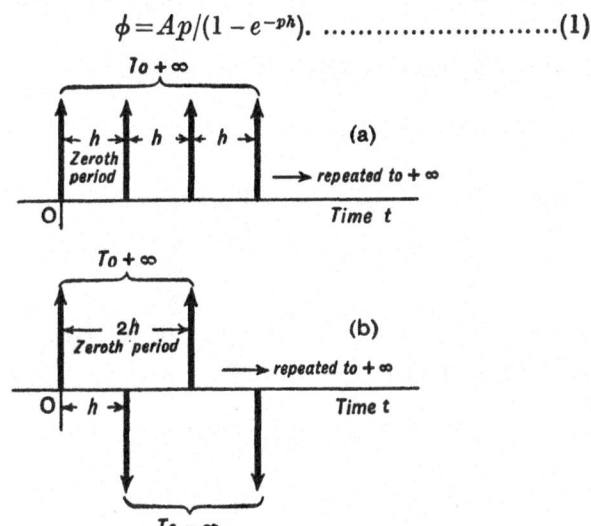

Fig. 24 (a).—Illustrating an impulsive force of infinite amplitude, but finite strength, repeated at interval h.

(b).—As at (a) but alternate impulses reversed.

If alternate impulses are reversed (Fig. 24b), the L.T. of the zeroth period $t = 0$ to $2h$, is by (1) § 7·141,

$$\phi(0 \; ; \; 2h) = Ap(1 - e^{-ph}), \quad\dots\dots\dots\dots\dots\dots(2)$$

$- Ape^{-ph}$ being the L.T. of the reversed impulse at $t = h$. Thus by (3) § 7·141,

$$\phi = Ap/(1 + e^{-ph}). \qquad \qquad \text{(3)}$$

Neither (1) nor (3) can be interpreted as a Fourier expansion, because the contour integrals corresponding to (4) § 7·14 diverge. When, however, either of these L.T.S. is used on the r.h.s. of a transform equation like (2), § 3·11, and the l.h.s. has one or more terms of the type p^n, $n \geqslant 1$, the ϕ so obtained is invertible, being a Fourier expansion, since the corresponding contour integral is convergent.

7·18. Example. A heavy mass m rests on a frictionless horizontal plane. Commencing at $t = 0$, a series of unidirectional blows each of strength A is imparted to m at interval h, by means of a hammer. What is v the velocity of m at any time $t > 0$?

The differential equation of motion is

$$m\frac{dv}{dt} = \xi_1(t), \qquad \qquad \text{(1)}$$

where $\xi_1(t)$ represents the sequence of impulsive blows. By (10) § 2·14 and (1) § 7·17, the transform equation for (1) is

$$mp\phi = Ap/(1 - e^{-ph}), \qquad \qquad \text{(2)}$$

where $v(t) \rightleftharpoons \phi$, and A represents the strength of the impulse of infinite amplitude. Thus

$$\phi = (A/m)/(1 - e^{-ph}), \qquad \qquad \text{(3)}$$

and by (4) § 7·14

$$v = \frac{(A/m)}{2\pi i} \int_{c-i\infty}^{c+i\infty} e^{zt}\, dz/z(1 - e^{-zh}), \qquad \qquad \text{(4)}$$

$$= (A/m)\left[(t/h) + \left\{ \frac{1}{2} + \frac{1}{\pi} \sum_{n=1}^{\infty} \frac{\sin (2\pi nt/h)}{n} \right\} \right], \qquad \text{(5)}$$

on evaluating at the poles of the integrand in (4) [12, pp. 48–50]. Formula (5) represents the 'staircase' function which is depicted in Fig. 25a. This will be understood more readily by considering the expansion of (3), namely,

$$\phi = \frac{A}{m}(1 + e^{-ph} + e^{-2ph} + \ldots). \quad \ldots\ldots\ldots\ldots\ldots\ldots(6)$$

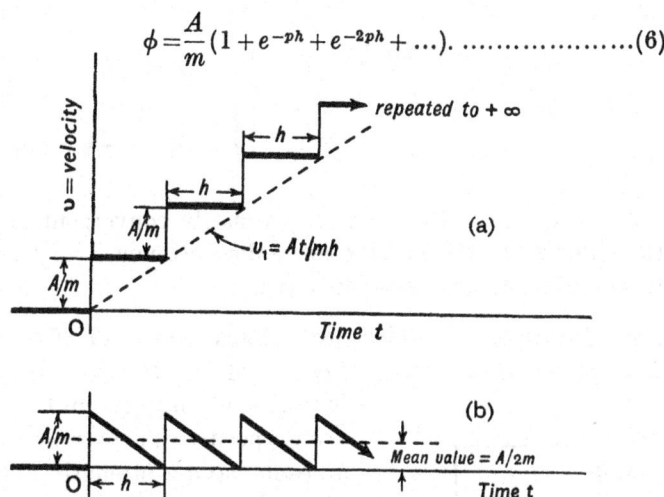

FIG. 25 (a).—The 'staircase' function, which is piecewise continuous, being finitely discontinuous at $t = nh$, n a positive integer, but unbounded at infinity.

(b).—The bounded 'saw-tooth' function which is periodic piecewise continuous in $t \geqslant 0$.

Term by term interpretation yields

$$v(t) = \frac{A}{m}[H(t) + H(t - h) + H(t - 2h) + \ldots]. \quad \ldots\ldots\ldots(7)$$

The first term is the unit function with amplitude A/m; the second is identical but commences at $t = h$; and so on. The superimposition of the unit functions results in an unending staircase. The equation of the straight line through the lower corners of the stairs is

$$v_1 = \frac{A}{m}\left(\frac{t}{h}\right), \quad \ldots\ldots\ldots\ldots\ldots\ldots\ldots\ldots\ldots(8)$$

i.e. the first term in (5). Removing that portion of the stairs above this line and plotting it as shown in Fig. 25b, we obtain a saw-tooth wave form, whose Fourier expansion is given by the part $(A/m) \{ \ \}$ in (5). Applying Dirichlet's test [3, 4] it can be shown that the series is u.c. in $0 < t < h$; $h < t < 2h$; etc. In these intervals the series represents the continuous functions $\left(\dfrac{1}{2}-\dfrac{t}{h}\right)$; $\left(\dfrac{3}{2}-\dfrac{t}{h}\right)$; etc., respectively. Thus

$$\frac{1}{\pi}\sum_{n=1}^{\infty}\frac{\sin 2\pi nt/h}{n}=\left(\frac{1}{2}-\frac{t}{h}\right),\quad 0<t<h, \quad\dots\dots\dots\dots(9)$$

$$=\left(\frac{3}{2}-\frac{t}{h}\right),\quad h<t<2h, \quad\dots\dots\dots(10)$$

and so on.

At $t=h$, $2h$, ... the series is ordinarily convergent and has the value zero, but the function is discontinuous. $A/2m$ is the mean value of the saw-tooth wave.

7·19. Example. An electrical impulse generator of resistance R is applied to the quiescent circuit of Fig. 26 as $t \to -0$. If the impulses, of the type in § 7·17, are always positive and occur at interval h, each having strength A (whose dimensions are voltage × time), find an expression for I_1, the current in inductance L.

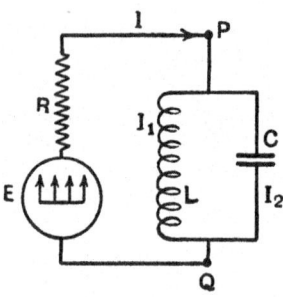

FIG. 26.—Schematic diagram of electrical circuit with impulse generator of resistance R.

The p.d. across PQ is

$$E-RI=L\frac{dI_1}{dt}=\frac{1}{C}\int_0^t I_2\,dt, \quad\dots(1)$$

whilst $I=I_1+I_2.$ $\quad\dots\dots\dots\dots\dots(2)$

The transform equations corresponding to (1), (2) are, by § 3·181 and (1) § 7·17,

$$\frac{Ap}{1-e^{-ph}}-R\phi=Lp\phi_1=\frac{\phi_2}{Cp}, \quad\dots\dots\dots\dots\dots(3)$$

$$\phi=\phi_1+\phi_2, \quad\dots\dots\dots\dots\dots\dots(4)$$

where $I \Rightarrow \phi$, $I_1 \Rightarrow \phi_1$, and $I_2 \Rightarrow \phi_2$. Solving (3), (4) taking $a=1/CR$, $b=1/LC$, $b>a^2/4$ (for an oscillatory circuit), we obtain

$$\phi_1=\frac{(A/LCR)p}{(p^2+ap+b)(1-e^{-ph})}, \quad\dots\dots\dots\dots\dots\dots(5)$$

so by Appendix IV

$$I_1=\frac{(A/LCR)}{2\pi i}\int_{c-i\infty}^{c+i\infty}\frac{e^{zt}\,dz}{(z^2+az+b)(1-e^{-zh})}. \quad\dots\dots\dots(6)$$

Since the integrand has 'simple' poles only, the condition for validity of the theorem of residues is satisfied (see below (7) § 7·16) and its application leads to

$$I_1 = \frac{A}{LCR}\left[\frac{e^{-\frac{1}{2}at}\sin kt - e^{-\frac{1}{2}a(t-h)}\sin k(t+h)}{k(1+e^{ah}-2e^{\frac{1}{2}ah}\cos kh)} + \frac{1}{bh} + \right.$$

$$\left. + 2h \sum_{n=1}^{\infty} \frac{\sin\left(2\pi nt/h + \tan^{-1}\dfrac{bh^2 - 4\pi^2n^2}{2\pi nah}\right)}{[(bh^2 - 4\pi^2n^2)^2 + 4\pi^2n^2a^2h^2]^{1/2}} \right] \quad ...(7)$$

where $k = \sqrt{b - a^2/4}$. The first member on the r.h.s. of (7) represents the transient which dies away rapidly. The term A/Rh represents the constant unidirectional current which increases with decrease in h and vice versa. This and the third member represents the Fourier expansion of the current in L, when t is large enough for the transient to be neglected. The Fourier expansion for the alternating current in L is given by the infinite series alone. By applying the M test, as in § 7·16, it may be shown that the series is absolutely and uniformly convergent for all t. Hence the current is a continuous function of t in the range $t \geqslant 0$. There is a sinusoidal current having the frequency of repetition of the impulses, namely, $1/h$. It is accompanied by an infinite series of harmonics whose frequencies are integral multiples of $1/h$ ($n = 2, 3, 4, ...$). Moreover, all frequencies are independent of the circuit constants. The amplitudes and phases of the currents are dependent, however, upon $a = 1/CR$ and $b = 1/LC$. The amplitude of the harmonic of frequency $2n/h$ is a maximum when

$$C = [h^2 + 4\pi^2n^2(L/R)^2]/4\pi^2n^2L. \quad ....(8)$$

If alternate impulses are reversed, the zeroth period being $t = (0, 2h)$, the current in L is

$$I_1 = \frac{A}{LCR}\left[\frac{e^{-\frac{1}{2}at}\sin kt + e^{-\frac{1}{2}a(t-h)}\sin k(t+h)}{k(1+e^{ah}+2e^{\frac{1}{2}ah}\cos kh)} - \right.$$

$$\left. -2h \sum_{n=0}^{\infty} \frac{\sin\{(2n+1)\pi t/h + \tan^{-1}[bh^2 - (2n+1)^2\pi^2]/[(2n+1)\pi ah]\}}{\{[bh^2 - (2n+1)^2\pi^2]^2 + (2n+1)^2\pi^2a^2h^2\}^{1/2}} \right]$$

$$(9)$$

In this case there is no unidirectional current, while all the harmonics are odd multiples of the repetition frequency, namely, $1/2h$. The amplitude of the harmonic of frequency $(2n + 1)/2h$ is a maximum when

$$C = [h^2 + (2n + 1)^2\pi^2(L/R)^2]/(2n + 1)^2\pi^2L. \quad\ldots\ldots\ldots(10)$$

The reader should verify all the results in this section as an exercise.

FOREWORD

THE appendices which follow are intended for readers who
may be unacquainted with (a) Heaviside's unit function,
(b) various definitions and theorems pertaining to infinite
series and infinite integrals. They should be helpful to the
technologist in particular, since his College curriculum is
usually conspicuous for its lack of these mathematical essentials.
To economise in space, proofs of the theorems are omitted, but
will be found in references [3, 4, 6, 7, 8, 16]. The subject
matter of these appendices is important in dealing with the
Laplace transform and its applications.

Many of the infinite series and infinite integrals encountered
in solving practical problems fulfil the necessary conditions for,
(a) term by term differentiation and integration, (b) differentiation and integration under the integral sign, (c) inversion of the
order of integration in a repeated infinite integral. This is
doubtless due to the physical nature of the problems solved.
A potential difference applied to one end of a long unloaded
submarine telegraph cable, causes a current to flow in the
cable. If a convergent series can be found to represent the
p.d. at any point of the cable, it *appears* extremely likely that
term by term differentiation of this series will lead to a formula
for the current. This is merely intuition, and not mathematical
proof. As stated in [ref. 8, p. 130] intuition in mathematics
is sometimes false. If a costly design is based upon intuitive
reasoning, the results may on occasion be disastrous. Accordingly it is inadvisable to trust to luck and make tacit assump-
tions regarding series and integrals. The appendices which
follow will enable the reader to ascertain whether or not
the conditions necessary for validity of various operations
mentioned above are satisfied. Moreover, it will be possible

to deal with the analysis mathematically and, therefore, logically, while intuition may serve as a guide.

It may be remarked that the conditions stipulated for convergence of series and integrals are usually *sufficient*. There are, however, cases where a condition may not be *necessary*. Instances are given in footnotes.

APPENDIX I

HEAVISIDE'S UNIT FUNCTION $H(t)$

This is illustrated graphically in Fig. 27a. It is zero when $t<0$, unity when $t>0$, but finitely discontinuous at $t=0$, where it has no definite value. If a function $f(t)$ is multiplied by $H(t)$, the part from $t=-\infty$ to -0 is suppressed, whilst that from $t=+0$ to $+\infty$ is retained. Thus

$$\cos t\, H(t)=\cos t\ \left.\begin{array}{l}0<t<+\infty, \quad \text{i.e. } t>0,\\ -\infty<t<0, \quad \text{i.e. } t<0.\end{array}\right\}\ \ \text{........(1)}$$

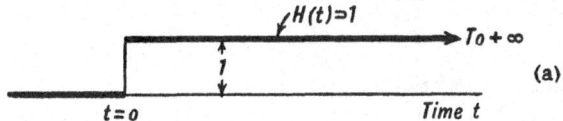

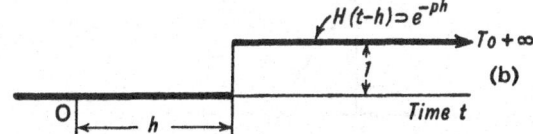

Fig. 27 (a) (b).—Illustrating Heaviside's unit function with origin at (a) $t=0$, (b) $t=h$, where it is finitely discontinuous and has no definite value. The thin vertical line (conventional) is not part of the graph.

When the origin of $H(t)$ is moved to the point $t=h>0$, we write

$$H(t-h)=1\ \left.\begin{array}{l}h<t<+\infty, \quad \text{i.e. } t>h,\\ -\infty<t<h, \quad \text{i.e. } t<h.\end{array}\right\}\ \ \text{...........(2)}$$

This is illustrated in Fig. 27b. Application to the Bessel function of the first kind of order zero gives

$$J_0(t)H(t-h)=J_0(t)\ \left.\begin{array}{l}t>h,\\ t<h.\end{array}\right\}\ \ \text{.................(3)}$$

Here $J_0(t)$ is obliterated from $-\infty$ to $(h-0)$. If we write $J_0(t-h)H(t-h)$, the origin is moved to $t=h$ for both functions, and the Bessel function commences at $t=h$ with the value

$J_0(0) = 1$. These two cases are illustrated in Figs. 28a and 28b, whilst $\cos(t - h) H(t - h)$ is depicted in Fig. 29.*

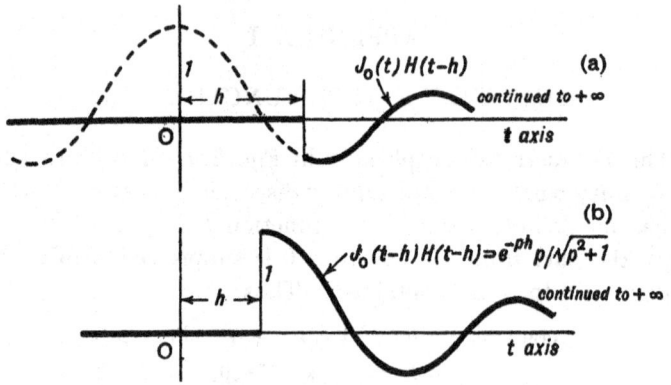

FIG. 28 (a).—The Bessel function

$$J_0(t) H(t - h) = \frac{p}{\sqrt{p^2 + 1}} - p \int_0^h e^{-pt} J_0(t) \, dt,$$

which is finitely discontinuous at $t = h$.

(b).—The B.F. $J_0(t - h) H(t - h) = e^{-ph} p / \sqrt{p^2 + 1}$, finitely discontinuous at $t = h$.

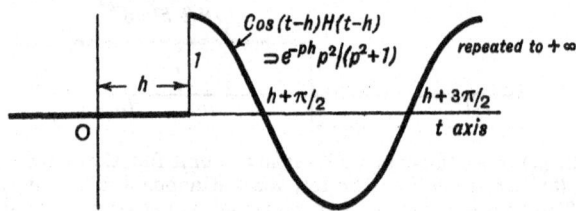

FIG. 29.—The circular function $\cos(t - h) H(t - h) = e^{-ph} p^2 / (p^2 + 1)$, which is finitely discontinuous at $t = h$.

By virtue of the discontinuity at $t = 0$, $H(t)$ cannot be differentiated there, although a number of writers on technical mathematics have made well-meaning attempts to do so. In applications the procedure employed frequently leads to correct results, but even mathematics is not free from flukes.

It may be remarked that $H(t) = 1$, and that $H(t - h) = e^{-ph}$.

* Additional examples will be found in reference 12, pp. 154, 155.

APPENDIX II

CONVERGENCE OF INFINITE SERIES

1. Let $u_0(t)$, $u_1(t)$, $\ldots u_n(t)$, $\ldots$, be functions of the real variable t, e.g. $u_n(t) = t^n/(2n)!$ Suppose that for any finite value of n we can represent the function $S(t)$ by a sum of $(n+1)$ terms of the form

$$S(t) = u_0(t) + u_1(t) + u_2(t) + \ldots + u_{n-1}(t) + R_n(t), \quad \ldots\ldots(1)$$
$$= S_n(t) + R_n(t). \quad \ldots\ldots\ldots\ldots\ldots\ldots\ldots\ldots\ldots\ldots\ldots\ldots\ldots\ldots(2)$$

Then if the infinite series $\sum\limits_{m=0}^{\infty} u_m(t)$ converges to $S(t)$, $R_n(t)$ is called the remainder after n terms. The *necessary and sufficient* condition* for convergence of the infinite series to $S(t)$ is evidently that

$$\lim_{n \to \infty} R_n(t) \to 0. \quad \ldots\ldots\ldots\ldots\ldots\ldots\ldots\ldots(3)$$

Alternatively, given any positive number ϵ, however small, we must be able to find a positive integer N such that

$$|R_n(t)| < \epsilon, \quad \ldots\ldots\ldots\ldots\ldots\ldots\ldots\ldots(4)$$

whenever $n > N$. The integer N (made definite by being chosen as small as possible) will usually depend upon t as well as upon ϵ.

2. A series which does not converge is said to diverge. It may diverge definitely to $+\infty$ or to $-\infty$, or it may oscillate over a finite† or an infinite range. For instance the well-known geometric series

$$S(t) = \sum_{n=0}^{\infty} t^n = 1 + t + t^2 + \ldots + t^n + \ldots , \quad \ldots\ldots\ldots\ldots(1)$$

---

* A condition may sometimes be *sufficient* but not *necessary* to ensure convergence, and vice versa.

† A series which oscillates over a finite range (constant or variable) is sometimes considered to be merely oscillatory and not divergent. It is certainly not convergent, since it does not tend to a definite limit as $n \to \infty$.

converges to the value $1/(1-t)$, provided that $-1 < t < 1$. When $t = 1$, $S_n(t) = n \to \infty$ and, therefore, the infinite series diverges to $+\infty$. But if $t = -1$, the terms are alternately $+1$ and -1, and the infinite series oscillates between 0 and 1.

3. The series $\sum\limits_{n=0}^{\infty} u_n(t)$ is said to be *absolutely* convergent if $\sum\limits_{n=0}^{\infty} |u_n(t)|$ is convergent, i.e. when all the terms are made positive. For example if $t > 0$, the series

$$S(t) = 1 - t + \frac{t^2}{2!} - \frac{t^3}{3!} + \ldots = e^{-t}, \quad \ldots\ldots\ldots\ldots(1)$$

is absolutely convergent, since it converges to e^t when all its terms are made positive. If $\sum\limits_{n=0}^{\infty} |u_n(t)|$ is convergent, then $\sum\limits_{n=0}^{\infty} u_n(t)$ is certainly convergent. The proof is left to the reader.

4. Particular convergence tests.

1°. A series of *positive* terms $\sum\limits_{n=0}^{\infty} u_n$

converges if $\qquad \lim\limits_{n \to \infty} \dfrac{u_n}{u_{n+1}} > 1,$

but diverges if $\qquad \lim\limits_{n \to \infty} \dfrac{u_n}{u_{n+1}} < 1.$

2°. A series of *positive* terms $\sum\limits_{n=0}^{\infty} u_n$

converges if $\qquad \lim\limits_{n \to \infty} n \left(\dfrac{u_n}{u_{n+1}} - 1 \right) > 1,$

but diverges if $\qquad \lim\limits_{n \to \infty} n \left(\dfrac{u_n}{u_{n+1}} - 1 \right) < 1.$

3°. When u_n is positive, the series $\sum\limits_{n=0}^{\infty} (-1)^n u_n$ converges

if $\quad (a) \qquad u_n > u_{n+1},$

and $(b) \qquad u_n \to 0$ as $n \to \infty$.

Thus a series of decreasing terms, alternately $+$ and $-$, which tend to zero as a limit, is convergent. It should be observed that the three tests above are applicable whether or not u_n is a function of t.

5. Examples. 1°. Show that the series for sinh t converges for all finite t. When $t = 0$, the convergence is obvious, so we take $t \neq 0$. Now

$$\sinh t = t + \frac{t^3}{3!} + \frac{t^5}{5!} + \ldots + \frac{t^{2n+1}}{(2n+1)!} + \ldots , \qquad \ldots\ldots(1)$$

so

$$\frac{u_n}{u_{n+1}} = \frac{t^{2n+1}}{(2n+1)!} \cdot \frac{(2n+3)!}{t^{2n+3}} = \frac{(2n+2)(2n+3)}{t^2} \to \infty \quad \ldots\ldots(2)$$

as $n \to \infty$, and consequently the series converges for all finite t.

2°. Discuss convergence of the hypergeometric series

$$_2F_1(\alpha, \beta; \gamma; t) = 1 + \frac{\alpha\beta t}{1!\gamma} + \frac{\alpha(\alpha+1)\beta(\beta+1)t^2}{2!\gamma(\gamma+1)} + \ldots , \quad \ldots(1)$$

when $t = 1$, α, β, γ real. Here

$$\frac{u_n^*}{u_{n+1}} = \frac{(n+1)(n+\gamma)}{(n+\alpha)(n+\beta)} \to 1 \text{ as } n \to \infty , \qquad \ldots\ldots\ldots\ldots(2)$$

so test 1° § 4 fails. But

$$n\left(\frac{u_n}{u_{n+1}} - 1\right) = \frac{n^2(\gamma - \alpha - \beta + 1) + n(\gamma - \alpha\beta)}{n^2 + n(\alpha + \beta) + \alpha\beta} , \qquad \ldots\ldots\ldots(3)$$

$$= \frac{(\gamma - \alpha - \beta + 1) + (\gamma - \alpha\beta)/n}{1 + (\alpha + \beta)/n + \alpha\beta/n^2} , \qquad \ldots\ldots\ldots\ldots(4)$$

$$\to \gamma - \alpha - \beta + 1 \text{ as } n \to \infty . \qquad \ldots\ldots\ldots\ldots\ldots(5)$$

Hence when $t = 1$, the series converges if $\gamma - \alpha - \beta + 1 > 1$, or $\gamma > \alpha + \beta$. When $0 < t < 1$,

$$u_n/u_{n+1} = [(n+1)(n+\gamma)/(n+\alpha)(n+\beta)t] > 1 \text{ when } n \to \infty ,$$

so the series converges unconditionally.

6. Uniform Convergence. 1°. If the series $u_0(t) + u_1(t) + \ldots$ converges in the *closed*† interval $h_1 \leqslant t \leqslant h$ to the sum $S(t)$, we may write

$$S(t) = S_n(t) + R_n(t), \qquad \ldots\ldots\ldots\ldots\ldots\ldots(1)$$

where $S_n(t)$ is the sum of the first n terms, and $R_n(t)$ the re-

* For convenience the $(n+2)$th term has been taken for u_n.

† In some cases uniform convergence may apply to the infinite range $t \geqslant 0$, while for convenience in §§ 7·15, 7·18 an open interval has been used. It is then to be understood that the convergence ceases to be uniform when the variable is arbitrarily near to each end of the interval.

mainder after *n terms*. We saw in § 1 above that we can assign
a positive number ϵ, however small, and *then* choose a positive
integer N, such that $|R_n(t)| < \epsilon$, whenever $n > N$. The integer
N usually depends upon t as well as upon ϵ. But if we can
choose N *independent* of t, we say that the series is *uniformly*
convergent. When this is so, the function $S(t)$ and the series

$\sum\limits_{n=0}^{\infty} u_n(t)$ have the same definite value at each point of the closed
interval $h_1 \leqslant t \leqslant h$, including the end points h_1, h, and the function is *continuous* throughout the interval. In fact the sum of

a *uniformly* convergent series of continuous functions $\sum\limits_{n=0}^{\infty} u_n(t)$,
is a continuous function [reference 16, p. 7]. The point to be
emphasised is that a u.c. series *always* represents a continuous
function. But we must not fall into the error of supposing the
converse to be true. A c.f. may in certain cases be represented
by a non-uniformly convergent series [ref. 16, p. 12].

   Example. The series $\sum\limits_{n=0}^{\infty} te^{-nt}$ converges in *any* finite interval
$0 \leqslant t \leqslant h$. When $t=0$, the series is zero for all values of n, so
$S_n(0) = S(0) = 0$, this being the value to which the series converges. If $t > 0$, we may write

$$S_n(t) = t(1 + e^{-t} + e^{-2t} + \ldots + e^{-(n-1)t}) = t(1 - e^{-nt})/(1 - e^{-t}). \ldots(2)$$

When $n \to \infty$ and $t > 0$, $e^{-nt} \to 0$, so we have the function

$$S(t) = t/(1 - e^{-t}). \ldots\ldots\ldots\ldots\ldots\ldots(3)$$

For t small, $(1 - e^{-t}) \simeq 1 - 1 + t = t$, so that as $t \to 0$, $S(t) \to 1$.* Thus
at the origin the series is zero, but the function is unity, so a
discontinuity occurs. Moreover, owing to its being non-
uniformly convergent as $t \to +0$, the series does not represent the
function *there*. It represents the function for $t \geqslant t_1$, $t_1 > 0$. The
convergence may be examined as follows :

$$R_n(t) = S(t) - S_n(t) = te^{-nt}/(1 - e^{-t}) \simeq e^{-nt}, \ldots\ldots\ldots\ldots(4)$$

if t is small. When n is large but finite, $R_n(t)$ depends upon t.
If we select ϵ arbitrarily small, we cannot make $e^{-nt} < \epsilon$ in
the neighbourhood of $t=0$, unless N is a function of t, e.g.

---

*Strictly $S(t)$ has no value at $t=0$, but we have used the 'limit.'

$n > N = -\log \epsilon/t$. Thus the series $\sum\limits_{n=0}^{\infty} te^{-nt}$ although convergent in $0 \leqslant t \leqslant h$, *does not converge uniformly* if t is arbitrarily near to zero. It is u.c., however, in any closed interval $h_1 \leqslant t \leqslant h, h_1 > 0$. This may be proved by aid of 2°.

2°. 'M' *test for uniform convergence.* The series $\sum\limits_{n=0}^{\infty} u_n(t)$ is absolutely and uniformly convergent in the closed interval $h_1 \leqslant t \leqslant h$, if a convergent series of *positive constant* terms $\sum\limits_{n=0}^{\infty} M_n$ (independent of t) can be found such that $|u_n(t)| \leqslant M_n$ *for every value of t in the interval.*

3°. *Example.* The exponential series $1 - t + \dfrac{t^2}{2!} - \dfrac{t^3}{3!} + \dots = e^{-t}$

is absolutely and uniformly convergent in *any* closed interval $h_1 \leqslant t \leqslant h$. For if H be the greater of $|h_1|$ and $|h|$, we have $0 \leqslant |t| \leqslant H$ for all t in the interval, so $|t^n/n!| \leqslant H^n/n!$ If we write $M_n = H^n/n!$, the series of positive constant terms $\sum\limits_{n=0}^{\infty} M_n$ is convergent, since $M_n/M_{n+1} = (n+1)/H \to +\infty$ with n. Hence the exponential series is absolutely and uniformly convergent in any closed interval.

4°. *Example of non-uniform convergence.* The series

$$1 + t + \frac{t^2}{2!} + \frac{t^3}{3!} + \dots + \frac{t^n}{n!} + R_{n+1}(t) = e^t$$

is u.c. in $h_1 \leqslant t \leqslant h$, but not in the range $t \geqslant 0$. Here $t^n/n! < R_{n+1}(t)$,[*] and for u.c. if ϵ is fixed, we must have $t^n/n! < R_{n+1}(t) < \epsilon$ for *all* $t \geqslant 0$, when $n >$ some positive number N. For this inequality to be satisfied, we must take N large enough to make $N! > t^N/\epsilon$, so N depends upon t when the latter is arbitrarily large. Hence the exponential series is not u.c. as $t \to +\infty$. In like manner it can be shown that the series for e^{-t} is not u.c. as $t \to -\infty$. Additional tests for u.c. will be found in references [3, 4].

[*] Provided t is large enough for a given n, since

$$R_{n+1}(t) = \sum\limits_{r=1}^{\infty} \frac{t^{n+r}}{(n+r)!} = \frac{t^n}{n!}\left[\frac{t}{n+1} + \frac{t^2}{(n+1)(n+2)} + \dots\right], \text{ and } \frac{t}{n+1} > 1.$$

7. Differentiation of Infinite Series Term by Term.

If $S(t) = \sum\limits_{n=0}^{\infty} u_n(t)$, converges in $a \leqslant x \leqslant b$, or in $x \geqslant 0$,

then $S'(t) = \sum\limits_{n=0}^{\infty} u_n'(t)$,

i.e. the derivative of the sum is equal to the sum of the derivatives, provided that in the above ranges of x,

(i) $\sum\limits_{n=0}^{\infty} u_n'(t)$ is uniformly convergent,

(ii) each $u_n'(t)$ is continuous.

APPENDIX III

CONVERGENCE OF INFINITE INTEGRALS

I. 1°. *Definition.* Assume that the integral $\int_{h_1}^{h} f(t)\, dt$ exists* for every finite value of $h > h_1$. This usually implies that $f(t)$ is either continuous, piecewise continuous, or has an infinite discontinuity of order < 1 (see § 1·211) in $h_1 \leqslant t \leqslant h$. As $h \to +\infty$ this integral may

$\qquad$ (a) tend to a finite limit,

$\qquad$ (b) $\quad$,, $\quad +\infty$,

$\qquad$ (c) $\quad$,, $\quad -\infty$,

$\qquad$ (d) oscillate over a finite range,

$\qquad$ (e) $\quad$,, $\quad$,, an infinite range.

In case (a) we say that $\int_{h_1}^{\infty} f(t)\, dt$ is convergent, and we write

$$\int_{h_1}^{\infty} f(t)\, dt = \lim_{h \to +\infty} \int_{h_1}^{h} f(t)\, dt. \quad\text{.....................(1)}$$

In all other cases we say that the l.h.s. of (1) is divergent, and the integrals do not 'exist'. Throughout this Appendix, except in § 2, we shall assume that all integrals exist in any finite range $h_0 \leqslant t \leqslant h_1$, which includes the lower limit. The various tests refer to convergence as the upper limit $h \to +\infty$. Cases where the integrand has an infinity at the origin are treated in § 1·211, and in § 2 below.

---

* An integral may be considered to 'exist' when its value—which may be zero—is finite and *definite*. 2° (d) is an example of a finite, indefinite integral. If $h_1 = 0$ and $f(t) = t^{-3/2}$, an infinite discontinuity of order 3/2 occurs at the origin, and for all h in $h > h_1$, the value of the integral $\to +\infty$ as $h_1 \to +0$, so it does not exist. If, however, $f(t) = t^{-1/2}$ the O.I. at the origin is 1/2, and the integral exists for all finite $h > 0$. It $\to +\infty$ with h. A divergent integral does not exist.

2°. *Examples.*

(a) $\displaystyle\int_1^\infty dt/t^2 = \lim_{h\to+\infty}\int_1^h dt/t^2 = 1\,;$

(b) $\displaystyle\lim_{h\to+\infty}\int_0^h t\,dt = +\infty\,;$

(c) $\displaystyle\lim_{h\to+\infty}\int_0^h (1-t)\,dt = -\infty\,;$

(d) $\displaystyle\int_0^h \cos t\,dt = \sin h,$

but $\sin h$ does not tend to a limit as $h\to+\infty$, so $\displaystyle\int_0^\infty \cos t\,dt$ oscillates finitely, i.e. it is bounded and *indefinite* ;

(e) $\displaystyle\int_0^h t\cos t\,dt = h\sin h + \cos h - 1,$

which oscillates over an infinite range as $h\to+\infty$, being unbounded and indefinite.

2. Improper integral ; Principal value.

If $|f(t)|\to+\infty$ as $t\to h_1+0$, $\displaystyle\lim_{\epsilon\to+0}\int_{h_1+\epsilon}^h f(t)\,dt$ may exist. It is written $\displaystyle\int_{h_1}^h f(t)\,dt$, and is termed an improper integral. For instance the three integrals in 1°–3° § 1·211 are in this category. All the integrands have infinities at the origin.

If $|f(t)|\to+\infty$ as $t\to t_1$, where $h_1 < t_1 < h$, then

$$\lim_{\epsilon\to+0}\int_{h_1}^{t_1-\epsilon} f(t)\,dt + \lim_{\epsilon_1\to+0}\int_{t_1+\epsilon_1}^h f(t)\,dt$$

may exist, and it is written $\displaystyle\int_{h_1}^h f(t)\,dt$. When ϵ, $\epsilon_1\to+0$ independently, neither limit may exist. Nevertheless

$$\lim_{\epsilon\to+0}\left[\int_{h_1}^{t_1-\epsilon} f(t)\,dt + \int_{t_1+\epsilon}^h f(t)\,dt\right]$$

may exist. It is known as Cauchy's principal value of $\displaystyle\int_{h_1}^h f(t)\,dt$, and is usually symbolised by $P\displaystyle\int_{h_1}^h f(t)\,dt$. $\displaystyle\int_0^2 dt/(t-1)$ is a case in point. Its value is not $\Big[\log(t-1)\Big]_0^2$, for this would be

$-\log(-1) = -\pi i$, which is impossible, since the integrand is real in the range $t = (0, 2)$. We have

$$\lim_{\epsilon \to +0} \left[\int_0^{1-\epsilon} dt/(t-1) + \int_{1+\epsilon}^2 dt/(t-1) \right]$$

$$= \operatorname{im} \left\{ \Big[\log(t-1) \Big]_0^{1-\epsilon} + \Big[\log(t-1) \Big]_{1+\epsilon}^2 \right\}$$

$$= \lim_{\epsilon \to +0} \left\{ \log(-\epsilon) - \log(-1) + \log 1 - \log \epsilon \right\},$$

so $P \int_0^2 dt/(t-1) = \lim_{\epsilon \to +0} (\log \epsilon + \log 1 - \log \epsilon) = 0$. A rough graph of $1/(t-1)$ for the range $t = (0, 2)$ will make this result clear. The graph is anti-symmetrical about the asymptote at $t = 1$, so the areas between the curves and the t axis are equal but of opposite sign.

3. General test for convergence of infinite integral.

$1°$. The integral $\int_{h_1}^\infty g(t)\,dt$ is said to converge if, *after* choosing a positive number ϵ, however small, $h > h_1*$ can be found such that

$$\left| \int_h^l g(t)\,dt \right| < \epsilon, \quad \ldots\ldots\ldots\ldots\ldots(1)$$

where l may have any finite value $> h$, or may $\to +\infty$. If the integral

$$\int_{h_1}^\infty |g(t)|\,dt \quad \ldots\ldots\ldots\ldots\ldots\ldots(2)$$

converges, then $\int_{h_1}^\infty g(t)\,dt$ is said to converge absolutely. If (2) converges, so also does the latter integral, but the converse is not necessarily true.

$2°$. *Using the test in* $1°$, *show that* $\int_0^\infty e^{-t}\,dt$ *is convergent.*

(a) $\int_0^h e^{-t}\,dt$ exists and has the value $1 - e^{-h}$, however large h (finite);

(b) $\left| \int_h^l e^{-t}\,dt \right| = e^{-h} - e^{-l} < e^{-h}.$ $\ldots\ldots\ldots\ldots\ldots\ldots(1)$

Choose any positive real quantity ϵ, however small, e.g. 10^{-20}. Then $e^{-h} < \epsilon$ if h is large enough but finite. Hence the original

* As stated in § 1, the integral is assumed to exist in any finite range $h_0 \leqslant t \leqslant h_1$.

integral converges. Integral (1) represents the area A between e^{-t} and the t axis over the range $t=h$ to l. Thus in effect, the condition for convergence is that A must be finite as $l \to +\infty$.

$3°$. *Using $1°$ show that $\int_0^\infty t^{-1/2}\,dt$ is divergent.*

(a) $\int_0^h t^{-1/2}\,dt$ exists for all finite $h>0$ and has the value $2\sqrt{h}$.

(b) $\left| \int_h^l t^{-1/2}\,dt \right| = 2\,(l^{1/2} - h^{1/2})$. ....(1)

Choose any positive real ϵ, e.g. 10^{-3}. Then however large h may be, l may be chosen $> h$ so that $2\,(l^{1/2} - h^{1/2}) > \epsilon$. Hence the integral diverges. As $l \to +\infty$, so also does the area A between $t^{-1/2}$ and the t axis from h to l.

It should be noticed that although h may be as large as we please, provided it is finite, l may exceed h by any number we please. For instance, if h were 10^{20}, l might be 10^{50} or $\to +\infty$.

4. Particular tests for convergence of infinite integrals. The general test in § 3 is sometimes inconvenient to apply, so two particular tests are given below. They enable convergence to be investigated without evaluating the integrals. In this respect it may be remarked that although an integral may be evaluable, this does not in itself constitute a *formal proof* of convergence. Additional tests will be found in § 7 of this Appendix.

$1°$. If (a) $\int_{h_1}^\infty f(t)\,dt$ converges,

(b) $g(t)$ is a bounded monotonic function* in the range $t \geqslant h_1$,

---

* A function which either steadily increases (monotonic increasing or m.i.) or decreases (m.d.) to a limiting value without oscillation. $A\,(1 - e^{-pt})$ and erf t [Fig. 10] are positive and m.i. to the respective limits A and unity, as $t \to +\infty$. $A\,(1 + e^{-pt})$ and erfc t [Fig. 10] are positive and monotonic decreasing to the respective limits A and zero, as $t \to +\infty$. Further examples will be found in Figs. 3, 4, 14, 15, 16, 17. A function of the type $t^n e^{-t}$ satisfies (b) in $1°$, $2°$, since it is bounded at its maximum value $n^n e^{-n}$ at $t=n$, and is m.d. when $t>n$. If $p = u + iv$, $e^{-pt} = e^{-ut}(\cos v - i\sin v)$. $e^{-ut}\cos v$ and $e^{-ut}\sin v$ are damped alternating functions which $\to 0$ as $t \to +\infty$. The envelope e^{-ut} is monotonic decreasing, but the function e^{-pt} is *not* monotonic. A function is sometimes said to be monotonic if its derivative does not change sign.

then $\displaystyle\int_{h_1}^{\infty} f(t)g(t)\,dt$ is convergent.

2°. If (a) $\displaystyle\int_{h_1}^{h} f(t)\,dt$ oscillates between finite limits, i.e. it is

bounded, as $h \to +\infty$,

and (b) $g(t)$ is a bounded monotonic function in $t \geqslant h_1$
which $\to 0$ as $t \to +\infty$,

then $\displaystyle\int_{h_1}^{\infty} f(t)g(t)\,dt$ is convergent.

In 1°, 2°, $\displaystyle\int_{h_1}^{\infty} g(t)\,dt$ need not be convergent, e.g. $\displaystyle\int_{1}^{h} dt/t = \log h$,
which $\to +\infty$ with h. Nevertheless if (a) is satisfied the theorems are valid. In (a) 2°, the integral must oscillate between *finite* upper and lower limits. For instance if
$f(t) = 1 + \cos t$, then $\displaystyle\int_{0}^{h} f(t)\,dt = (h + \sin h)$ whose graph is $\sin h$
superimposed on a straight line, through the origin, at 45° to the h axis. Although $\sin h$ oscillates between finite limits, the limits of $(h + \sin h) \to +\infty$ with h, so the theorem is invalid here.

The term 'bounded' is introduced to ensure that $g(t)$ has no infinity in the range of integration. There are cases, however, when the integral converges provided the order of infinity of $g(t)$ is less than unity (see § 1·211). For instance $\displaystyle\int_{0}^{\infty} \sin t\, dt/t^{\nu}$ is
convergent for $0 < \nu < 2$ (see footnote to 4° § 7). When $0 < \nu \leqslant 1$ there is no infinity, but when $1 < \nu < 2$ there is an infinity of order < 1 at the origin. If $\nu = 2$, the O.I. is unity and the integral diverges.

5. Examples.   1°. Show that $\displaystyle\int_{1}^{\infty} \frac{\sin t}{t} \tanh t\, dt$ is convergent.

By 2° § 4, $\displaystyle\int_{1}^{\infty} \sin t\, dt/t$ is convergent, since $\displaystyle\int_{1}^{h} \sin t\, dt$ oscillates
between the finite limits $\pm 1 + \cos 1$ as $h \to +\infty$, whilst in the range $t \geqslant 1$, $1/t$ is a bounded monotonic decreasing function which $\to 0$ as $t \to +\infty$. Also $\tanh t$ is a monotonic increasing

function with upper bound unity.   Hence by 1° § 4—with $f(t) = \sin t/t$ and $g(t) = \tanh t$—the original integral is convergent.

2°. Discuss the convergence of $\displaystyle\int_1^\infty \cos t\, dt/t^q$, q real; (a) $q > 1$; (b) $0 < q \leqslant 1$; (c) $q = 0$; (d) $q < 0$.

(a) $|\cos t| \leqslant 1$, so $\displaystyle\int_1^\infty \left|\frac{\cos t}{t^q}\right| dt \leqslant \int_1^\infty \frac{dt}{t^q} = 1/(q-1),$ ......(1)

provided $q > 1$, so the original integral is absolutely convergent.

(b) $\displaystyle\int_1^h \cos t\, dt$ oscillates finitely as $h \to +\infty$, whilst if $0 < q \leqslant 1$, t^{-q} is monotonic decreasing and $\to 0$ as $t \to +\infty$. Hence by 2° § 4 the original integral converges.   (c) If $q = 0$, we get $\displaystyle\int_1^h \cos t\, dt$ which oscillates finitely as $h \to +\infty$, whilst (d) if $q < 0$ the range of oscillation $\to \pm\infty$ as $h \to +\infty$.   Thus the integral diverges if $q \leqslant 0$.

3°. *Show that* $\displaystyle\int_0^\infty [\mathbf{H}_0(at) - Y_0(at)] \cos bt\, dt$ *is convergent, a real* > 0, *b real* > 0.

Divide the range of integration into a finite part $t = (0, h)$ and an infinite part $t = (h, +\infty)$. Since $\mathbf{H}_0(at)$ is a bounded damped alternating function [see ref. 11, Fig. 13], $\displaystyle\int_0^h \mathbf{H}_0(at) \cos bt\, dt$ exists. When at is small, $Y_0(at) = \mathbf{O}^*(\log at)$, so if h_1 is small, $\displaystyle\int_0^{h_1} Y_0(at) \cos bt\, dt$ converges by comparison with $\displaystyle\int_0^{h_1} \mathbf{O}(\log at)\, dt = \mathbf{O}[h_1(\log ah_1 - 1)]$ which is bounded.   Since $Y_0(at)$ is a bounded damped alternating function when $at \geqslant h_1$, $h_1 > 0$, it follows that the original integral exists in any finite range $t = (0, h)$.   For large values of t, we may use the asymptotic formula [ref. 11, p. 71]

$$\mathbf{H}_0(at) - Y_0(at) \sim (2/\pi at) - \mathbf{O}(1/t^3). \quad\text{...............}(1)$$

* See p. xii for meaning of this symbol.

Thus for the infinite range $t = (h, +\infty)$, we have

$$\int_h^\infty [\mathbf{H}_0(at) - Y_0(at)] \cos bt \, dt$$

$$\sim \frac{2}{\pi a} \int_h^\infty \frac{\cos bt}{t} \, dt - \int_h^\infty O\left(\frac{\cos bt}{t^3}\right) dt. \quad \ldots\ldots(2)$$

By 2° § 4 both integrals on the r.h.s. of (2) are convergent. Hence the original integral is convergent.

4°. *Under what conditions does* $\int_0^\infty e^{\pm i\omega t} f(t) \, dt$ *converge?* Since $e^{\pm i\omega t} = \cos \omega t \pm i \sin \omega t$, by 2° § 4 the integral converges if $f(t)$ is bounded and monotonic decreasing in $t \leqslant 0$ and $\to 0$ as $t \to +\infty$. Sometimes convergence occurs if $f(t)$ is a damped oscillatory function, e.g. $J_0(t)$, $J_0^2(t)$, the former being an alternating and the latter a positive d.o.f. When t is large enough

$$J_0(t) \sim \sqrt{2/\pi t} \cos (t - \tfrac{1}{4}\pi),$$

so it is possible by applying 2° § 4 to prove that

$$\int_h^\infty \begin{Bmatrix} \cos \omega t \\ \sin \omega t \end{Bmatrix} J_0(t) \, dt \quad \text{and} \quad \int_h^\infty \begin{Bmatrix} \cos \omega t \\ \sin \omega t \end{Bmatrix} J_0^2(t) \, dt$$

converge, with $h \gg 0$. In the first case we must have $\omega \neq 1$, and in the second $\omega \neq 2$. Since $\int_0^h$ exist, it follows that $\int_0^\infty$ are convergent. The reader should verify this as an exercise.

5°. *Show that the integral in* (2) § 5·151 *is convergent,* t *real* > 0.

Obviously the integral exists in the finite range $x = (0, h)$, so we shall consider the range $x = (h, +\infty)$. When h is large enough, the asymptotic formula for the Bessel function may be used, so with $a = 2\sqrt{t}$

$$\int_h^\infty J_0(a\sqrt{x}) \sin x \, dx$$

$$\sim \sqrt{\frac{2}{\pi a}} \int_h^\infty \cos(a\sqrt{x} - \tfrac{1}{4}\pi) \sin x \, dx/x^{1/4}, \quad \ldots\ldots\ldots(1)$$

$$= \frac{1}{\sqrt{\pi a}} \int_h^\infty (\cos a\sqrt{x} + \sin a\sqrt{x}) \sin x \, dx/x^{1/4}. \quad \ldots\ldots(2)$$

Writing $x = y^2$, we have $dx = 2y\,dy$, and (2) becomes

$$I = \frac{2}{\sqrt{\pi a}} \int_{\sqrt{h}}^{\infty} y^{1/2} (\cos ay + \sin ay) \sin y^2\,dy. \quad \ldots\ldots\ldots\ldots(3)$$

Now $\quad \sin y^2 \cos ay$
$$= \tfrac{1}{2} [\sin (y^2 + ay) + \sin (y^2 - ay)]$$
$$= \tfrac{1}{2} [\sin \{(y + \tfrac{1}{2}a)^2 - \tfrac{1}{4}a^2\} + \sin \{(y - \tfrac{1}{2}a)^2 - \tfrac{1}{4}a^2\}], \quad \ldots\ldots(4)$$

and $\quad \sin y^2 \sin ay$
$$= \tfrac{1}{2} [\cos \{(y - \tfrac{1}{2}a)^2 - \tfrac{1}{4}a^2\} - \cos \{(y + \tfrac{1}{2}a)^2 - \tfrac{1}{4}a^2\}]. \quad \ldots\ldots(5)$$

Let $\quad (y + \tfrac{1}{2}a)^2 - \tfrac{1}{4}a^2 = v$, then $2(y + \tfrac{1}{2}a)\,dy = dv$, so
$dy = dv/2v_1$ where $v_1 = \sqrt{v + \tfrac{1}{4}a^2}$; also $y^{1/2} = \sqrt{v_1 - \tfrac{1}{2}a}$.
Taking $(y - \tfrac{1}{2}a)^2 - \tfrac{1}{4}a^2 = u$ gives $dy = du/2u_1$, where $\quad \ldots\ldots(6)$
$u_1 = \sqrt{u + \tfrac{1}{4}a^2}$; also $y^{1/2} = \sqrt{u_1 + \tfrac{1}{2}a}$. When $y = \sqrt{h}$,
$v = (\sqrt{h} + \tfrac{1}{2}a)^2 - \tfrac{1}{4}a^2 = k$, and $u = (\sqrt{h} - \tfrac{1}{2}a)^2 - \tfrac{1}{4}a^2 = l$.

Substituting from (4)–(6) into (3), we obtain

$$I = \frac{1}{2\sqrt{\pi a}} \left[\int_k^{\infty} \frac{(v_1 - \tfrac{1}{2}a)^{1/2}}{v_1} (\sin v - \cos v)\,dv \right.$$
$$\left. + \int_l^{\infty} \frac{(u_1 + \tfrac{1}{2}a)^{1/2}}{u_1} (\sin u + \cos u)\,du \right]. \quad \ldots\ldots\ldots(7)$$

The non-circular functions in the integrands are bounded in $v \geqslant k$, $u \geqslant l$, monotonic and $\to 0$ as v and $u \to +\infty$. Thus by $2° \S 4$ it follows that the integrals converge. Hence the original integral does likewise. When k, l are adequately large

$$I \simeq \frac{1}{(2\pi)^{1/2} t^{1/4}} \int_h^{\infty} \frac{\sin v}{(v + t)^{1/4}}\,dv. \quad \ldots\ldots\ldots\ldots(8)$$

6. Uniform convergence of infinite integral.

$1°$. The integral $\int_{h_1}^{\infty} f(x, t)\,dt$ is said to converge* uniformly *for all values of x in the closed interval $a \leqslant x \leqslant b$,* if

(a) *after* choosing a positive number ϵ, however small, a positive number n can be found such that

$$\left| \int_h^l f(x, t)\,dt \right| < \epsilon, \quad \ldots\ldots\ldots\ldots\ldots(1)$$

where $l > h > n$, and

(b) n is *independent* of x in the interval, i.e. it is the same for

* In accordance with the statement in $\S 1$, we assume $\int_{h_1}^{h}$ to exist.

all values of x in the interval. The number n depends, of course, upon ϵ, i.e. $n = n(\epsilon)$.

When evaluated, $\int_{h_1}^{\infty} f(x, t)\,dt$ is a function of x and h_1, say, $g(x, h_1)$. If

(c) $f(x, t)$ is a continuous or a piecewise continuous function of x in the interval $a \leqslant x \leqslant b$, and

(d) the integral is *uniformly convergent* in the interval, then the function $g(x, h_1)$ is a *continuous* function of x in the interval, and represents the value of the integral for all x therein. The point to be emphasised is that *uniform convergence of the integral implies continuity of the function it represents.* The converse, however, is not necessarily true (see §§ 1·211, 1·22, 1·23, 1° § 8 this Appendix). Uniform convergence is usually considered in a closed interval with respect to the variable therein—x in the above case—but not with respect to the variable of integration—t in the above case. The infinite range $x \geqslant 0$, is sometimes employed.

2°. Using the test in 1°, show that $\int_0^{\infty} e^{-pt}\,dt$ is u.c. in any real interval $p_0 \leqslant p \leqslant p_1$, $p_0 > 0$.

(a) $\int_0^h e^{-pt}\,dt$ has the value $(1 - e^{-ph})/p$, for all finite $h > 0$;

(b) $\int_h^l e^{-pt}\,dt = (e^{-hp} - e^{-lp})/p < e^{-hp}/p.$

Choose any real positive number ϵ, however small. Then *whatever* the value of p in the above interval, e^{-hp}/p can be made less than ϵ by making h large enough and greater than some positive number n, itself dependent on ϵ. Hence the integral converges uniformly with respect to p in the interval $p_0 \leqslant p \leqslant p_1$, $p_0 > 0$, and the function it represents $(1/p)$ is continuous for all p in this interval. Introducing numerical values, let $\epsilon = 10^{-50}$, $p_0 = 10^{-10}$, then we must have $e^{-hp}/p_0 < \epsilon$ or $e^{hp_0} > 1/p_0\epsilon$. Taking logarithms to the base 10, the condition becomes

$$h > \log_{10}(1/p_0\epsilon)/p_0 \log_{10} e,^*$$

so $h > \log_{10} 10^{60}/10^{-10} \times 0\cdot4343 \ldots = 6 \times 10^{11}/0\cdot4343.$

* Taking logarithms to the base e, $h > - (\log p_0 + \log \epsilon)/p_0$, and h is finite if p_0 and ϵ are finite.

The requisite condition would certainly be satisfied for all $p_0 > 10^{-10}$ if $h > n = 1 \cdot 5 \times 10^{12}$, and although large, this number is finite.

7. Particular tests for uniform convergence of infinite integral.

1°. The 'M' test. If

    (a) in the interval $a \leqslant x \leqslant b$, $|f(x, t)| \leqslant M(t)$, where $M(t)$ is a *positive function independent of x*,*

    (b) $\displaystyle\int_{h_1}^{\infty} M(t)\,dt$ converges,

then $\displaystyle\int_{h}^{\infty} f(x, t)\,dt$ is absolutely and uniformly convergent for all x in $a \leqslant x \leqslant b$.

2°. If

    (a) $\displaystyle\int_{h_1}^{\infty} g(t)\,dt$ converges,

    (b) for every value of x in $a \leqslant x \leqslant b$, the function $f(x, t)$ is bounded, monotonic decreasing, and positive (t variable and $\geqslant h$),†

then $\displaystyle\int_{h_1}^{\infty} f(x, t)g(t)\,dt$ converges uniformly with respect to x, for all x in the closed interval $a \leqslant x \leqslant b$.

3°. By 2° it follows that if $\displaystyle\int_{0}^{\infty} f(t)\,dt$ is convergent, $\displaystyle\int_{0}^{\infty} e^{-pt}f(t)\,dt$ is uniformly convergent in the interval $0 \leqslant p \leqslant p_1$ for every *real* finite value of p_1. When p is complex, the integral is uniformly convergent if $R(p) > 0$ [see ref. 6, pp. 111–2, conditions II]. A simpler theorem states that if $\displaystyle\int_{0}^{\infty} e^{-pt}f(t)\,dt$ converges for $p = p_0$, it converges uniformly for all values of $p = u + iv$ in the finite part of the p plane where $u > u_0$. In other words it is u.c. for all values of p in a *closed* area where $R(p - p_0) > 0$.

4°. If (a) $\displaystyle\int_{h_1}^{h} g(t)\,dt$ oscillates between finite limits as $h \to +\infty$,

---

* The theorem is true also if $|f(x, t)| \leqslant M(t)$ in the range $x \geqslant 0$.

† $f(x, t)$ may be constant for an assigned value of x, e.g. $e^{-xt} = 1$, when $x = 0$, or like te^{-xt} it may be monotonic increasing to its maximum value $(xe)^{-1}$ at $t = 1/x$, after which it is monotonic decreasing to zero as $t \to +\infty$.

(b) $f(x, t)$ is m.d.* and positive for every x in $a \leqslant x \leqslant b$
(t variable $\geqslant h_1$),

(c) $f(x, t) \to 0$ uniformly with respect to every x in
$a \leqslant x \leqslant b$ as $t \to +\infty$,

then $\displaystyle\int_{h_1}^{\infty} f(x, t) g(t) dt$ converges uniformly with respect to x,
for all x in $a \leqslant x \leqslant b$.

8. Examples. 1°. Show that $\displaystyle\int_0^{\infty} \frac{\sin xt \, dt}{a^2 + t^2}$ converges uniformly
with respect to x in any finite interval thereof.

Using 1° § 7 we have $f(x, t) = \sin xt/(a^2 + t^2)$, and

$$\left| \frac{\sin xt}{a^2 + t^2} \right| \leqslant \frac{1}{(a^2 + t^2)}$$

throughout any finite interval of variation of x. Also

$$\int_0^h \frac{dt}{a^2 + t^2} = \frac{1}{a} \left[\tan^{-1} t/a \right]_0^h \to \pi/2a$$

as $h \to +\infty$, so this integral converges. Hence the original
integral is absolutely and uniformly convergent. It can also be
proved u.c. by 4° § 7. This is left as an exercise for the reader.

2°. Show that $\displaystyle\int_0^{\infty} e^{-pt} J_\nu(t) dt$ is *uniformly* convergent if
p real $\geqslant 0$, $R(\nu) > -1$.

Using 2° § 7, we take $g(t) = J_\nu(t)$, and we have to demonstrate
that the integral $\displaystyle\int_0^{\infty} J_\nu(t) dt$ converges. We write this integral in
the form

$$\int_0^{\infty} J_\nu(t) dt = \int_0^h J_\nu(t) dt + \int_h^{\infty} J_\nu(t) dt, \quad \dots\dots\dots(1)$$

h being large but finite. The first integral on the r.h.s. has a
definite finite value only if $R(\nu) > -1$. For [ref. 11, p. 158]

$$J_\nu(t) = \frac{(\tfrac{1}{2}t)^\nu}{\Gamma(1 + \nu)} + \text{terms involving higher powers of } t, \dots(2)$$

* See footnote to 1° (b) § 4. In some cases $f(x, t)$ may have an infinite
discontinuity. For instance $\int_0^{\infty} x \sin at \, dt/t^\nu = \pi x a^{\nu-1} \operatorname{cosec} \left(\dfrac{\nu\pi}{2}\right) \Big/ 2\Gamma(\nu)$,
which holds for $0 < \nu < 2$. When $1 < \nu < 2$, the factor $1/t^\nu$ introduces an
infinity at $t = 0$, but its order being $(\nu - 1)$ is less than unity (see § 1·211).

so the above restriction is needed to secure convergence at $t=0$. In the second integral on the r.h.s. of (1), we use the asymptotic formula [11, p. 158],

$$J_\nu(t) \sim \sqrt{(2/\pi t)} \cos (t - \tfrac{1}{4}\pi - \tfrac{1}{2}\nu\pi) + \zeta(t), \quad \ldots\ldots\ldots(3)$$

where $\zeta(t) = \mathbf{O}(t^{-3/2})$,* when t is large enough. Since

$$\int_h^\infty \zeta(t)\,dt = \int_h^\infty \mathbf{O}(t^{-3/2})\,dt \quad \ldots\ldots\ldots\ldots\ldots(4)$$

is convergent, we need consider the integral

$$\sqrt{(2/\pi)} \int_h^\infty \cos (t - \tfrac{1}{4}\pi - \tfrac{1}{2}\nu\pi)\,dt/t^{1/2} \quad \ldots\ldots\ldots\ldots(5)$$

only. By $2° \,\S\,4$ this converges, so the integral on the l.h.s. of (1) is convergent. If p real >0, $f(p, t)=e^{-pt}$ is m.d. positive and $\leqslant 1$ for $t \geqslant 0$; for $p=0$ it is constant (see footnote to (b) $2° \,\S\,7$). Hence by $2° \,\S\,7$, the integral $\int_0^\infty e^{-pt}J_\nu(t)\,dt$ is *uniformly* convergent if p real $\geqslant 0$ and $R(\nu) > -1$.

9. Theorems.—Changing the order of integration.

A. *One upper limit b finite, the other h infinite.*

If (a) $f(x, t)$† is a continuous function of x in the interval $a \leqslant x \leqslant b$, and of t in the interval $h_1 \leqslant t \leqslant h$, no matter how large h may be,

(b) $\int_{h_1}^\infty f(x, t)\,dt$ converges *uniformly* with respect to x in the interval $a \leqslant x \leqslant b$,

then $$\int_{h_1}^\infty dt \int_a^b f(x, t)\,dx = \int_a^b dx \left[\int_{h_1}^\infty f(x, t)\,dt \right]. \quad \ldots\ldots\ldots\ldots(1)$$

By virtue of *uniform* convergence in (b) the *integral* there represents a continuous function of x, and is, consequently, integrable over the range $a \leqslant x \leqslant b$. Thus conditions (a), (b) imply

---

* See p. xii for explanation of this symbol.

† $f(x, t)$ can always be integrated with respect to x, if it is continuous in a finite range of integration. A similar remark applies in regard to t. When the range is infinite, continuity in itself is insufficient to secure convergence.

that the r.h.s. of (1) exists.* Also, since $f(x, t)$ is continuous in x, it follows that the finite integral $\int_a^b f(x, t)dx$ represents a continuous function of t,† and can, therefore, be integrated over the finite range $h_1 \leqslant t \leqslant h$. Moreover, the theorem states that $\lim\limits_{h \to +\infty} \int_{h_1}^h dt \int_a^b f(x, t)dx$ exists and has a definite finite value, which is equal to $\int_a^b dx \int_{h_1}^\infty f(x, t)dt$.

Upper limits in both integrals infinite. It is difficult to specify general conditions to be satisfied here. In theorems B, C, the conditions are sufficient, but not always necessary. When the integrals are evaluable both before and after inversion, with identical results, no formal proof is usually needed. For a detailed discussion of the subject, the reader may consult the references in [4, p. 211].

B. *Both upper limits infinite.*

If (a) $f(x, t)$ is a *positive* continuous function of x in $a \leqslant x \leqslant b$, and of t in $h_1 \leqslant t \leqslant h$, no matter how large b and h may be,

$$(b) \quad \int_{h_1}^\infty dt \int_a^b f(x, t)dx = \int_a^b dx \int_{h_1}^\infty f(x, t)dt \quad \dots\dots\dots\dots(1)$$

for all values of $b > a$,

$$(c) \quad \int_{h_1}^h dt \int_a^\infty f(x, t)dx = \int_a^\infty dx \int_{h_1}^h f(x, t)dt \quad \dots\dots\dots\dots(2)$$

for all values of $h > h_1$,

then

$$\int_{h_1}^\infty dt \int_a^\infty f(x, t)dx = \int_a^\infty dx \int_{h_1}^\infty f(x, t)dt, \quad \dots\dots\dots\dots(3)$$

provided that one of these expressions has a meaning. Let $\lim\limits_{h \to \infty} \int_{h_1}^h dt \int_a^\infty f(x, t)dx$ have a definite finite value. Then (3)

* Let $\int_{h_1}^\infty f(x, t)dt = g(x, h_1)$, then by u.c. this function is continuous. Hence the finite integral $\int_a^b g(x, h_1)dx$, i.e. the r.h.s. of (1), exists.

† This function must fulfil certain conditions for the l.h.s. of (1) to converge.

asserts that $\lim\limits_{b\to\infty}\int_a^b dx \int_{h_1}^{\infty} f(x,\,t)dt$ has a definite finite value, and that these two values are equal.

C. Both upper limits infinite.

If (a) $f(x,\,t)$ is a continuous function of x in $a\leqslant x\leqslant b$, and of t in $h_1\leqslant t\leqslant h$, no matter how large b and h may be,

(b) $\int_a^{\infty} f(x,\,t)dx$ converges uniformly with respect to t in the arbitrary closed interval $h_1\leqslant t\leqslant h_2$,

(c) $\int_{h_1}^{\infty} f(x,\,t)dt$ converges uniformly with respect to x in the arbitrary closed interval $a\leqslant x\leqslant b_2$,

(d) $\int_a^{\infty} dx \int_{h_1}^{t} f(x,\,\tau)d\tau$ converges uniformly with respect to t in the infinite interval $t\geqslant h_1$,

then $\int_a^{\infty} dx \int_{h_1}^{\infty} f(x,\,t)dt$; $\int_{h_1}^{\infty} dt \int_a^{\infty} f(x,\,t)dx$ converge, and are equal.

C_1. *Sufficient conditions for the changed order of integration in* §§ 2·21, 2·22. These may be deduced from C above.

If (a) $f(t)$ is continuous in $t\geqslant 0$;

(b) $\int_{h_1}^{\infty} e^{-pt}f(t)\,dt$ is u.c. as a function of p in the *arbitrary* * real interval $p_1\leqslant p\leqslant p_2$;

(c) $\int_{p_1}^{\infty} e^{-pt}f(t)\,dp$ is u.c. as a function of t in the *arbitrary* * interval $h_1\leqslant t\leqslant h_2$;

(d) $\int_{h_1}^{\infty} f(t)\,dt \int_{p_1}^{p_2} e^{-pt}\,dp$ is u.c. as a function of p in the infinite range $p_2\geqslant p_1$; †

then $\int_{p_1}^{\infty} dp \int_{h_1}^{\infty} e^{-pt}f(t)\,dt = \int_{h_1}^{\infty} f(t)\,dt \int_{p_1}^{\infty} e^{-pt}\,dp$

$$= \int_{h_1}^{\infty} e^{-p_1 t}\frac{f(t)}{t}\,dt. \quad\ldots\ldots\ldots\ldots\ldots(1)$$

* This means that p or t may have *any* real finite value compatible with (b) or (c) being satisfied, respectively.

† p_1 may have any real finite value compatible with (d) being satisfied.

Since $\displaystyle\int_{p_1}^{p_2} e^{-pt} f(t)\, dt \leqslant e^{-p_1 t}\,\frac{|f(t)|}{t}$,

if $\displaystyle\int_{h_1}^{\infty} e^{-pt}\,\frac{|f(t)|}{t}\, dt$ converges, then by 1°, § 7, it follows

that $\displaystyle\int_{h_1}^{\infty} f(t)\, dt \int_{p_1}^{p_2} e^{-pt}\, dp$ is u.c. as a function of p in $p_2 \geqslant p_1$.

Hence instead of condition (d), we may use the less general but more easily applied condition, that

(d') $\displaystyle\int_0^{\infty} e^{-p_1 t}\,\frac{|f(t)|}{t}\, dt$ converges.

C_2. *Sufficient conditions for the changed order of integration in* § 2·23.

If (a) $f(at)$ is continuous in $t \geqslant 0$, and in $a \geqslant 1$;

(b) $\displaystyle\int_0^{\infty} e^{-pt}\frac{f(at)}{a}\, dt$ is u.c. as a function of a in $1 \leqslant a \leqslant b$;

(c) $\displaystyle\int_1^{\infty} e^{-pt}\frac{f(at)}{a}\, da$ is u.c. as a function of t in $0 \leqslant t \leqslant h$;

(d) $\displaystyle\int_1^{\infty} \frac{da}{a}\int_0^t e^{-pt} f(a\tau)\, d\tau$ is u.c. in $t \geqslant 0$;

then $\displaystyle\int_1^{\infty}\frac{da}{a}\int_0^{\infty} e^{-pt} f(at)\, dt = \int_0^{\infty} e^{-pt}\, dt \int_1^{\infty}\frac{f(at)}{a}\, da.$ ....(1)

10. Examples. 1°. *Show that*

$$\int_{p_1}^{\infty} dp \int_{h_1}^{\infty} e^{-pt}\sin t\, dt = \int_{h_1}^{\infty}\sin t\, dt \int_{p_1}^{\infty} e^{-pt}\, dp, \quad p_1 > 0,\ h_1 > 0.$$

We apply § 9, Theorem C, and,

(a) $e^{-pt}\sin t$ is a continuous function of p in $p \geqslant 0$, and of t in $t \geqslant 0$;

(b) $\displaystyle\int_{p_1}^{\infty} e^{-pt}\sin t\, dp = e^{-p_1 t}\frac{\sin t}{t}$, and converges uniformly to this value for all t in $h_1 \leqslant t \leqslant h_2$; also it is u.c. in this interval when the lower limit in the integral is 0 ;

(c) $\int_{h_1}^{\infty} e^{-pt} \sin t \, dt = \frac{e^{-ph_1}}{p^2+1} (p \sin h_1 + \cos h_1)$, and converges
uniformly to this value in $p_1 \leqslant p \leqslant p_2, \ p_1 > 0$;

(d) $\int_{p_1}^{\infty} dp \int_{h_1}^{t} e^{-p\tau} \sin \tau \, d\tau$

$$= \int_{p_1}^{\infty} \frac{dp}{(p^2+1)} [K^* - e^{-pt}(p \sin t + \cos t)]$$

and the convergence of this integral is independent of t in $t \geqslant 0$. Hence the two repeated
integrals converge and are equal, i.e. the order
of integration in either may be inverted. The
case where $p_1 = h_1 = 0$ is treated in 2°.

2°. *Prove that*

$$\int_0^{\infty} \sin t \, dt \int_0^{\infty} e^{-pt} dp = \int_0^{\infty} dp \int_0^{\infty} e^{-pt} \sin t \, dt.$$

By theorem § 9A

$$\int_0^{\infty} \sin t \, dt \int_{\delta}^{\Delta} e^{-pt} dp = \int_{\delta}^{\Delta} dp \int_0^{\infty} e^{-pt} \sin t \, dt, \ \dots\dots(1)$$

since (a) $e^{-pt} \sin t$ is a continuous function of p in $\delta \leqslant p \leqslant \Delta$,
and of t in $t \geqslant 0$;

(b) $\int_0^{\infty} e^{-pt} \sin t \, dt$ is u.c. with respect to p in $\delta \leqslant p \leqslant \Delta, \ \delta > 0$.

Consequently it is sufficient for our purpose to demonstrate that

$$\lim_{\delta \to +0} \int_0^{\infty} \sin t \, dt \int_0^{\delta} e^{-pt} dp = \lim_{\delta \to +0} \int_0^{\infty} \frac{\sin t}{t} (1 - e^{-\delta t}) \, dt = 0, \ \dots.(2)$$

and $\lim_{\Delta \to +\infty} \int_0^{\infty} \sin t \, dt \int_{\Delta}^{\infty} e^{-pt} dp = \lim_{\Delta \to +\infty} \int_0^{\infty} \frac{\sin t}{t} e^{-\Delta t} \, dt = 0. \ \dots..(3)$

By 2°, § 4, if $h > 0$, the individual integrals in

$$I = \int_h^{\infty} \frac{\sin t}{t} (1 - e^{-\delta t}) \, dt \ \dots\dots\dots\dots\dots(4)$$

converge. Suppose the value of the first is K, then that of the
second will be $- K e^{-\delta h_1}, \ 0 < h_1 < h$. Hence

$$I = K(1 - e^{-\delta h_1}) \to 0 \ \dots\dots\dots\dots\dots\dots(5)$$

$*K = e^{-ph_1}(p \sin h_1 + \cos h_1)$.

as $\delta \to +0$.   Also

$$\int_0^h \frac{\sin t}{t}(1 - e^{-\delta t})dt = \delta \int_0^h \sin t \left[1 - \frac{\delta t}{2!} + \ldots\right]dt \to 0, \quad \ldots\ldots(6)$$

as $\delta \to +0$.

Further $\qquad \int_0^\infty \frac{\sin t}{t} e^{-\Delta t} dt < \int_0^\infty e^{-\Delta t} dt = \frac{1}{\Delta} \to 0,$

as $\Delta \to +\infty$.   Hence (1) is true if $\delta \to +0$ and $\Delta \to +\infty$, i.e. the order of integration in the repeated integral may be inverted.

3°. *Evaluate* $p \displaystyle\int_0^\infty dx \int_0^\infty e^{-pt - x^2/4t} x t^{-1/2} dt$, *by inverting the order of integration, and justify this step.*

We have

$$p \int_0^\infty e^{-pt} t^{-1/2} dt \int_0^\infty e^{-x^2/4t} x \, dx$$

$$= 2p \int_0^\infty e^{-pt} t^{1/2} dt \int_0^\infty e^{-x^2/4t} d(x^2/4t), \quad \ldots\ldots(1)$$

$$= 2p \int_0^\infty e^{-pt} t^{1/2} dt = \sqrt{\pi/p}, \quad \ldots\ldots\ldots\ldots(2)$$

by the list.   To justify the changed order of integration, we proceed thus :

(a) $f(x, t) = e^{-pt - x^2/4t} x t^{-1/2}$, is a positive continuous function of x in the range $x \geqslant 0$, and of t in $t \geqslant 0$.   Thus condition (a) in § 9B is satisfied.

(b) For condition (b) § 9B to hold, (b) § 9A must be satisfied; so we have to prove that $\displaystyle\int_0^\infty e^{-pt - x^2/4t} x t^{-1/2} dt$ converges uniformly with respect to x in the range $x \geqslant 0$.   When

$x = 0$ and $\to +\infty$, $g(x, t) = x t^{-1/2} e^{-x^2/4t} = 0$ for all $t \geqslant 0$.

For the range $0 < x < \infty$, $g(x, t)$ is a maximum when $x = \sqrt{2t}$, and has the value $\sqrt{2/e}$.   Thus if p is real and > 0, in the range $x \geqslant 0$, we have

$$\left| e^{-pt - x^2/4t} x t^{-1/2} \right| \leqslant \sqrt{2/e} \cdot e^{-pt}. \quad \ldots\ldots\ldots\ldots(3)$$

Now $\sqrt{2/e} \displaystyle\int_0^\infty e^{-pt} dt$ converges to the value $\sqrt{2/ep^2}$, so by 1° § 7 the original integral converges uniformly in $x \geqslant 0$, and, therefore, represents a continuous function of x in this range.

(c) For condition (c) § 9B to hold, condition (b) § 9A must be satisfied, so we have to prove that the integral

$$\int_0^\infty e^{-pt-x^2/4t} x\, t^{-1/2}\, dx$$

converges uniformly with respect to t in the infinite range $t \geqslant 0$, when p is real and > 0.

To investigate the convergence we apply (1) § 6 making $l \to +\infty$. Then

$$2t^{1/2}e^{-pt}\int_{x=h}^{x\to\infty} e^{-x^2/4t}\, d\,(x^2/4t)$$

$$= 2t^{1/2}e^{-pt}\left[e^{-x^2/4t}\right]_{x\to\infty}^{x=h} = \frac{2t^{1/2}e^{-pt}}{e^{h^2/4t}}, \quad \ldots\ldots(4]$$

$$< 2e^{-pt}t^{1/2}/(h^2/4t) = 8e^{-pt}t^{3/2}/h^2, \quad \ldots\ldots\ldots(5)$$

If we select a real positive number ϵ, say 10^{-20}, by taking h large enough, but finite, $8t^{3/2}e^{-pt}/h^2 < \epsilon$ for all t in $t \geqslant 0$. For $t = 0$ we take the integrand as its limit zero as $t \to +0$. Hence the original integral is u.c. in the range $t \geqslant 0$. Since all conditions for the changed order of integration are fulfilled, (1) is valid and (2) is the value of the repeated integral.

11. Integration under the integral sign.

This operation is equivalent to changing the order of integration, so the conditions in § 9 are *sufficient* but not always *necessary* [see ref. 3, 16].

12. Theorems.—Differentiation under the integral sign.

A. Both limits in the integral finite.

If (a) $f(x, t)$ is a continuous function of x in $a \leqslant x \leqslant b$, and of t in $h_1 \leqslant t \leqslant h$,

(b) $\dfrac{\partial f(x,t)}{\partial x}$ is a continuous function of x in $a \leqslant x \leqslant b$, and of t in $h_1 \leqslant t \leqslant h$,

(c) h_1 and h are either constants, or functions of x $(a \leqslant x \leqslant b)$ with continuous differential coefficients, then

$$\frac{\partial}{\partial x}\int_{h_1}^h f(x, t)\,dt = f(x, h)\frac{dh}{dx} - f(x, h_1)\frac{dh_1}{dx} + \int_{h_1}^h \frac{\partial f(x, t)}{\partial x}\,dt. \quad \ldots(1)$$

When h, h_1 are constants, the first and second members on the r.h.s. of (1) vanish.

B. Upper limit of integral infinite.

If (a) $f(x, t)$ is a continuous function of x in $a \leqslant x \leqslant b$, and of t in $h_1 \leqslant t \leqslant h$, no matter how large h may be,

(b) $\dfrac{\partial f(x, t)}{\partial x}$ is a continuous function of x in $a \leqslant x \leqslant b$, and of t in $h_1 \leqslant t \leqslant h$, no matter how large h may be,

(c) $\displaystyle\int_{h_1}^{\infty} \dfrac{\partial f(x, t)}{\partial x} dt$ converges uniformly with respect to x, in $a \leqslant x \leqslant b$,

(d) $\displaystyle\int_{h_1}^{\infty} f(x, t) dt$ converges in $a \leqslant x \leqslant b$,

(e) h_1 is either constant or a function of x ($a \leqslant x \leqslant b$) having a continuous differential coefficient, then

$$\frac{\partial}{\partial x} \int_{h_1}^{\infty} f(x, t) dt = \int_{h_1}^{\infty} \frac{\partial f(x, t)}{\partial x} dt - f(x, h_1) \frac{dh_1}{dx}. \quad \ldots\ldots\ldots(1)$$

When h_1 is constant the last member of (1) vanishes.

13. Examples.

1°. Evaluate $\displaystyle\int_{b}^{\infty} e^{-pt} I_1(a\sqrt{t^2 - b^2})\, dt/\sqrt{t^2 - b^2}$ by differentiating $\displaystyle\int_{b}^{\infty} e^{-pt} I_0(a\sqrt{t^2 - b^2})\, dt$ under the integral sign with respect to b, given that $I_0(a\sqrt{t^2 - b^2}) \Rightarrow \dfrac{pe^{-b\sqrt{p^2 - a^2}}}{\sqrt{p^2 - a^2}}$.

Using (1) § 12B, with b instead of x, we have

$$\int_{h_1}^{\infty} \frac{\partial f(b, t)}{\partial b} dt = \int_{b}^{\infty} e^{-pt} \frac{\partial}{\partial b} I_0(a\sqrt{t^2 - b^2}) dt$$

$$= \int_{b}^{\infty} e^{-pt} \frac{\partial I_0(a\sqrt{t^2 - b^2})}{\partial (a\sqrt{t^2 - b^2})} \cdot \frac{\partial (a\sqrt{t^2 - b^2})}{\partial b} \cdot dt, \ldots\ldots\ldots(1)$$

$$= - \int_{b}^{\infty} e^{-pt} I_0{}'(a\sqrt{t^2 - b^2}) ab\, dt/\sqrt{t^2 - b^2}$$

$$= - ab \int_{b}^{\infty} e^{-pt} I_1(a\sqrt{t^2 - b^2})\, dt/\sqrt{t^2 - b^2}, \ldots\ldots\ldots(2)$$

since $I_0{}'(u) = I_1(u)$.   Also with $h_1 = b$,

$$f(b, h_1)\frac{dh_1}{db} = \{e^{-pt}I_0(a\sqrt{t^2-b^2})\}_{t=b}\frac{d(b)}{db} = e^{-pb}, \quad\ldots\ldots(3)$$

since $I_0(0) = 1$.   Hence by (2), (3), and (1) § 12B, we obtain

$$\frac{\partial}{\partial b}\int_b^\infty e^{-pt}I_0(a\sqrt{t^2-b^2})\,dt$$

$$= -ab\int_b^\infty \frac{e^{-pt}I_1(a\sqrt{t^2-b^2})}{\sqrt{t^2-b^2}}\,dt - e^{-pb}. \quad\ldots\ldots(4)$$

Now $$I = \int_b^\infty e^{-pt}I_0(a\sqrt{t^2-b^2})\,dt = e^{-b\sqrt{p^2-a^2}}/\sqrt{p^2-a^2}, \quad\ldots\ldots(5)$$

so $$\frac{dI}{db} = -e^{-b\sqrt{p^2-a^2}}. \quad\ldots\ldots\ldots\ldots\ldots(6)$$

Consequently by (4) and (6) we obtain the result

$$\int_b^\infty e^{-pt}\frac{I_1(a\sqrt{t^2-b^2})}{\sqrt{t^2-b^2}}\,dt = \frac{1}{ab}(e^{-b\sqrt{p^2-a^2}} - e^{-bp}). \quad\ldots\ldots(7)$$

The reader should justify the foregoing procedure as an exercise.

2°. Given that $\int_0^\infty e^{-pt}\sin xt\,dt = x/(p^2+x^2)$, p real and > 0, evaluate the integral $\int_0^\infty e^{-pt}t\cos xt\,dt$, by differentiating under the integral sign and justify this procedure.   Here

(a) $f(x, t) = e^{-pt}\sin xt$, and this is a continuous function of x and t in the respective ranges $a \leqslant x \leqslant b$, $0 \leqslant t \leqslant h$, no matter how large h may be, p real > 0 ;

(b) $\dfrac{\partial f(x, t)}{\partial x} = e^{-pt}t\cos xt$, and this is a continuous function of x and t in the respective ranges $a \leqslant x \leqslant b$, $0 \leqslant t \leqslant h$, no matter how large h may be, p real > 0 ;

(c) $\int_0^\infty \dfrac{\partial f(x, t)}{\partial x}\,dt = \int_0^\infty e^{-pt}t\cos xt\,dt$, and we have to examine whether or not this integral converges *uniformly* with respect to x in any finite range thereof.   Using 1° § 7 we see that $|e^{-pt}t\cos xt| \leqslant e^{-pt}t$, for any finite value of x, and $\int_0^\infty e^{-pt}t\,dt$ converges to the value $1/p^2$ (by (1) § 1·12).   Hence by 1° § 7 the integral in question converges uniformly with respect to x in $a \leqslant x \leqslant b$;

(d) the original integral converges in $a \leqslant x \leqslant b$, p real > 0.

Accordingly by § 12B

$$\frac{\partial}{\partial x} \int_0^\infty e^{-pt} \sin xt \, dt = \int_0^\infty e^{-pt} \frac{\partial}{\partial x} (\sin xt) dt = \frac{d}{dx} [x/(p^2 + x^2)], \quad ...(1)$$

so

$$\int_0^\infty e^{-pt} t \cos xt \, dt = (p^2 - x^2)/(p^2 + x^2)^2. \quad ....(2)$$

Since the integral is *u.c.* in $a \leqslant x \leqslant b$, the function on the r.h.s. of (2) is *continuous* in this range of x, i.e. for any p real > 0, it is a continuous function of x.

14. Example where (c) § 12B is not satisfied. The integral

$\int_0^\infty \sin xt \, \dfrac{dt}{t} = \tfrac{1}{2}\pi$, $x > 0$ is well known (see § 2·221). Differentiating under the integral sign with respect to x, we have

$$\int_0^\infty \frac{\partial}{\partial x} \left(\frac{\sin xt}{t} \right) dt = \int_0^\infty \cos xt \, dt, \quad ....(1)$$

but this operation is invalid, since the second member of (1) is oscillatory, instead of being *uniformly* convergent.

15. Theorem.—Integration of infinite series over an infinite range, term by term.

If (a) $\sum\limits_{n=0}^{\infty} f_n(t)$ is a uniformly convergent series* of continuous functions in the infinite range $t \geqslant h_1$;

(b) $\int_{h_1}^\infty | g(t) | \, dt$ is convergent,

then $\int_{h_1}^\infty g(t) \left[\sum\limits_{n=0}^{\infty} f_n(t) \right] dt = \sum\limits_{n=0}^{\infty} \int_{h_1}^\infty g(t) f_n(t) \, dt. \quad ....(1)$

Alternatively if

(c) $\sum\limits_{n=0}^{\infty} f_n(t)$ is a uniformly convergent series of continuous functions in the finite interval $h_1 \leqslant t \leqslant h$,

* There are non-uniformly convergent series which may be integrated term by term [see reference 16, § 1·75].

(d) $g(t)$ is a continuous function for all finite $t \geqslant h_1$,

(e) either the integral $\displaystyle\int_{h_1}^{\infty} |g(t)| \left\{ \sum_{n=0}^{\infty} |f_n(t)| \right\} dt,$ ....(2)

or the series $\displaystyle\sum_{n=0}^{\infty} \int_{h_1}^{\infty} |g(t)| \cdot |f_n(t)| \, dt,$ ....(3)

converges, then (1) is valid. This is exemplified in §§ 2·18, 2·181.

PARTICULAR CASE OF THE MELLIN INVERSION THEOREM

There are in reality two theorems as follows:

1°. If
$$\phi(p) = p \int_0^\infty e^{-pt} f(t)\, dt, \quad t \text{ real} > 0, \quad \ldots\ldots\ldots\ldots(1)$$

then
$$f(t) = \frac{1}{2\pi i} \int_{c-i\infty}^{c+i\infty} e^{zt} \phi(z)\, dz/z, \ldots\ldots\ldots\ldots\ldots\ldots(2)^*$$

provided

(a) All singularities of $\phi(z)/z$ are on the left of $c \pm i\infty$, $c > 0$;

(b) $\displaystyle\int_{-\infty}^{\infty} \left| \frac{\phi(c+iy)}{c+iy} \right| dy$ converges, but this restriction may be waived when the highest power in the expansion of $\phi(z)$ is z^ν, $R(\nu) < 1$;

(c) $|\phi(z)/z| \to 0$ uniformly with respect to phase z as $|z| \to \infty$ in the range $-\frac{1}{2}\pi \leqslant \text{phase } z \leqslant \frac{1}{2}\pi$. Integral (2) is then zero when $t < 0$.

2°. If
$$f(t) = \frac{1}{2\pi i} \int_{c-i\infty}^{c+i\infty} e^{zt} \phi(z)\, dz/z, \ldots\ldots\ldots\ldots\ldots(2a)$$

then
$$\phi(p) = p \int_0^\infty e^{-pt} f(t)\, dt, \ldots\ldots\ldots\ldots\ldots(1a)$$

provided

(a) $f(t)$ is continuous or piecewise continuous, t real > 0;

* See Fig. 18 for the contour $c \pm i\infty$, also [12]. Strictly $c \pm i\infty$ should be written $c \pm iy$, $y \to \infty$, i.e. integral (2) is the limit as $y \to \infty$. Although this theorem bears Mellin's name, it was given originally by Poisson in a memoir read before the Academy of Sciences, Paris, in 1815. Mellin gave a rigorous discussion of it in 1896, the form treated being that in [12, pp. 341–346]. For additional historical information see T. J. I'a Bromwich, *Proc. Lond. Math. Soc.*, **15**, 401, 1916; and H. Bateman, *Bull. Amer. Math. Soc.*, **48**, 510, 1942.

(b) $R(p) > 0$, but in some cases may be $\leqslant 0$;

(c)   integral (1a) is absolutely convergent.

When t is replaced by $a\sqrt{t^2 - b^2}$, a, t, b real, $t > b$, condition 1 (c) becomes : $\mid e^{zb}\phi(z)/z \mid \to 0$, etc. ; in (2a), t real $> b$; whilst in integrals (1), (1a) the lower limit is b. A complete proof of the theorem is given in reference 12, Appendix IV.

*APPENDIX V

PROOF THAT L.T. METHOD GIVES CORRECT SOLUTION OF ORDINARY LINEAR DIFFERENTIAL EQUATIONS WITH CONSTANT COEFFICIENTS

The transform solution of (1) § 3·13 given at (4) § 3·131 is

$$y(t) = y_p(t) + y_c(t) = \frac{\phi_1(p)}{\phi_2(p)} + \frac{\phi_3(p)}{\phi_2(p)}, \quad \ldots\ldots\ldots\ldots(1)$$

where $y_p = \phi_1/\phi_2$ is the solution for zero initial conditions, and $y_c = \phi_3/\phi_2$ is that part of the solution for the prescribed initial conditions. ϕ_1 is *usually* a rational function* such that the degree of the denominator of $\phi_1/p\phi_2$ exceeds that of the numerator by at least $(n+1)$.† Then the singularities of $\phi_1/z\phi_2$ are all poles in the finite part of the z plane, and the contour $c \pm i\infty$ may be replaced by a finite circle enclosing them. By the Mellin inversion theorem

$$y_p(t) = \frac{1}{2\pi i} \int_C e^{zt} \phi_1 \, dz/z\phi_2. \quad \ldots\ldots\ldots\ldots(2)$$

Since (2) and its first n derivatives are uniformly convergent in any closed interval $0 \leqslant t \leqslant h$, differentiation under the integral sign with respect to t is permissible, so

$$a_0 y_p^{(n)}(t) + a_1 y_p^{(n-1)}(t) + \ldots + a_n y_p(t)$$

$$= \frac{1}{2\pi i} \int_C \frac{e^{zt}\phi_1 \, dz}{z\phi_2} [a_0 z^n + a_1 z^{n-1} + \ldots + a_n], \quad \ldots\ldots(3)$$

$$= \frac{1}{2\pi i} \int_C e^{zt} \phi_1 \, dz/z = \xi(t). \quad \ldots\ldots\ldots\ldots\ldots(4)$$

The closed contour C is equivalent to $c \pm i\infty$, if ϕ_1 is a rational function whose degree $\leqslant 0$. It follows that the appropriate function $y_p(t)$ is obtained by inverting ϕ_1/ϕ_2.

*A rational function is one whose numerator and denominator are polynomials prime to each other. A constant is regarded as a polynomial.

† n is the order of the differential equation, and when p is large $\phi_1/p\phi_2 = O(1/p^{n+1})$. Since $t^{n+1} = (n+1)!/p^{n+1}$, it follows that y_p vanishes with t.

By (1), (2) § 3·131, ϕ_3 is a polynomial of degree equal to or less than that of ϕ_2, so applying the Mellin inversion theorem to the last member of (1), we have

$$y_c(t) = \frac{1}{2\pi i}\int_C e^{zt}\phi_3\,dz/z\phi_2, \quad\dotfill(5)$$

where C is equivalent to $c \pm i\infty$.

As in (2), (5) may be differentiated under the integral sign, so

$$a_0y_c^{(n)}(t) + a_1y_c^{(n-1)}(t) + \dots + a_ny_c(t) = \frac{1}{2\pi i}\int_C e^{zt}\phi_3\,dz/z. \quad\dots(6)$$

ϕ_3 has no constant term, so the integral corresponding to each term of the polynomial vanishes.   Hence if the initial conditions are satisfied, $y_c(t)$ is the initial conditions function of (1) § 3·13, i.e. if $y_p^{(r)}(0) + y_c^{(r)}(0) = y_r$, for $r = 0, 1, 2, \dots n-1$. By (2)

$$y_p^{(r)}(0) = \frac{1}{2\pi i}\left[\int_C e^{zt}z^r\phi_1\,dz/z\phi_2\right]_{t=0}, \quad\dotfill(7)$$

since all the integrals concerned are uniformly convergent in $0 \leqslant t \leqslant h$, and, therefore, represent $y_p^{(r)}(t)$ at $t = 0$.   Now $(n-1)$ is the greatest value of r, so $z^r\phi_1/z\phi_2 = \mathbf{O}\,(1/z^2)$,[*] and by reference 12 § 4·61 et seq., when t is small (7) may be written

$$y_p^{(r)}(t) \simeq \frac{1}{2\pi i}\int_C e^{zt}\,dz/z^m = \frac{t^{m-1}}{(m-1)!}, \quad\dotfill(8)$$

where m is an integer $\geqslant 2$.   Whence $y_p^{(r)}(0) = 0$.
By (5)

$$y_c^{(r)}(0) = \frac{1}{2\pi i}\left[\int_C e^{zt}z^r\phi_3\,dz/z\phi_2\right]_{t=0} = \frac{1}{2\pi i}\int_C z^r\phi_3\,dz/z\phi_2, \quad\dots(9)$$

since all the integrals concerned are uniformly convergent in $0 \leqslant t \leqslant h$, and, therefore, represent $y_c^{(r)}(t)$ at $t = 0$.

By aid of (1), (2) § 3·131, ϕ_3/ϕ_2 may be expressed as follows :

$$\phi_3/\phi_2 = \frac{1}{\phi_2}\{y_0(\phi_2 - a_n) + y_1[(\phi_2/z) - a_{n-1} - (a_n/z)]$$
$$+ y_2[(\phi_2/z^2) - a_{n-2} - (a_{n-1}/z) - (a_n/z^2)] + \dots\}, \quad\dots(10)$$

$$= y_0 + \frac{y_1}{z} + \frac{y_2}{z^2} + \dots + \frac{y_{n-1}}{z^{n-1}} - \frac{1}{\phi_2}\{(y_0a_n + y_1a_{n-1} + y_2a_{n-2} + \dots)$$
$$+ (1/z)[y_1a_n + y_2a_{n-1} + \dots] + \dots\}. \quad\dots(11)$$

[*] By hypothesis the degree of ϕ_1 never exceeds zero, while that of ϕ_2 is n.

When (11) is substituted in (9), the integrals corresponding to all terms except y_r/z^r vanish, so $y_c^{(r)}(0)=y_r$. Hence we have shown that

$$y_p^{(r)}(0)+y_c^{(r)}(0)=y_r, \quad \ldots\ldots\ldots\ldots\ldots(12)$$

for $r=0, 1, 2, \ldots n-1$, which completes the proof.

An exceptional case. The above analysis is based upon arbitrary initial conditions. If, however, $\xi(t)=I(t)$, the impulsive function of unit strength in § 2·19, $\phi_1=p$, so

$$z^r\phi_1/z\phi_2 = \mathbf{O}\,(z^{-1})$$

for $r=n-1$, and (7) gives

$$y_p^{(n-1)}(0)=a \text{ constant.}$$

Accordingly when a system symbolised by (1) § 3·13 is subjected to unit impulse, a ' forced ' initial condition is introduced by virtue of the $(n-1)$th derivative of y_p. This can, alternatively, be regarded as a condition at $t=+0$. It is due to the whole energy of the impulse being transferred instantaneously to the impulsed system. At $t=+0$, therefore, the initial condition is expressed by $y_p^{(n-1)}(0)+y_{n-1}$. It may be preferable to consider $y_0, y_1, \ldots y_{n-1}$ to represent the initial conditions as $t \to -0$, whilst at $t=0$, $y_p^{(n-1)}(0)$ represents a response of the system due to the impulse. See example 27 § 8·5, where $n-1=1$.

APPENDIX VI

SOLUTION OF D.E. FOR IMPULSIVE FUNCTION $I(t)$ WHEN THAT FOR UNIT FUNCTION IS KNOWN

For quiescent initial conditions, and a force represented by $H(t)$, applied at $t=0$, suppose that the solution of the differential equation for a given system is

$$f(t) \Rightarrow \phi(p). \qquad\qquad\qquad\qquad (1)$$

Then by (5) § 2·14, assuming the conditions there to be satisfied,

$$f'(t) \Rightarrow p\phi(p) - pf(0). \qquad\qquad\qquad (2)$$

Since $I(t) \Rightarrow p$, (2) can be written

$$A[f'(t) + I(t)f(0)] \Rightarrow Ap\,\phi(p), \qquad\qquad (3)$$

so the l.h.s. of (3) is the solution corresponding to $AI(t)$. The constant A is introduced for dimensional purposes, as explained in § 2·19. Terms in the original solution corresponding to nonzero initial conditions are independent of the applied force. They must be added to the l.h.s. of (3) to obtain the complete solution.

Variable $a\sqrt{t^2 - b^2} = au$. In this case the modified version of (3) is

$$A\left[\frac{d}{dt}f(au) + I(t-b)f(au)\right]_{t=b} \Rightarrow Ap\phi(p), \qquad\qquad (4)$$

where $\phi(p)$ is defined at (2) § 1·31. As before, terms arising from the initial conditions are to be added to the l.h.s. of (4) to obtain the complete solution (see 22, § 8·3).

Applying (4) to (6) § 4·311 (see (1) § 4·313), with $E_0 = 1$ we have

$$\frac{d}{dt}f(\beta u) = \sqrt{(\mathbf{C}/\mathbf{L})}\,\beta e^{-\alpha t}\left[t\,\frac{I_1(\beta u)}{u} - I_0(\beta u)\right], \qquad (5)$$

where $u = \sqrt{t^2 - x^2/v^2}$. Also

$$f(\beta u)_{t=x/v} = \sqrt{(\mathbf{C}/\mathbf{L})}e^{-\alpha x/v}. \qquad\qquad\qquad (6)$$

182

Hence by (4), (5), (6), if the p.d. applied to a quiescent transmission line at $x = 0$, when $t = 0$, is $AI(t)$, the current at any point x is given by

$$I_x = A\sqrt{(\mathbf{C}/\mathbf{L})}\left\{e^{-\alpha x/v}I(t - x/v) + \beta e^{-\alpha t}\left[t\frac{I_1(\beta u)}{u} - I_0(\beta u)\right]\right\}. \quad \dots(7)$$

This result may be obtained from (9) § 4·332 also, by writing $AI(t)$ for $\xi_2(t)$, and using (7) § 2·191. The value of the integral in (9) § 4·332 with $AI(t - \mu)$ substituted for $\xi_2(t - \mu)$ is found on replacing μ by t and removing the integral sign. The first term of (7), namely, $A\sqrt{(\mathbf{C}/\mathbf{L})}e^{-\alpha x/v}I(t - x/v)$ represents a current impulse at the point x at time $t = x/v$. This is followed by a 'tail' represented by the remaining terms of (7). It is due to dissipation of the energy, supplied instantaneously at $t = 0$, at a finite rate when $t > 0$. The current sinks to zero asymptotically.

PROBLEMS*

8·1. Boundedness, convergence of series, differentiation under the integral sign.

1. Show that (a) $e^{-p_0 t} f(t)$ is bounded in $t \geqslant 0$ when $f(t) = t^n$, $\sinh t$, $\cosh t$, $t^n \cosh t$, if p_0 real > 1 (p_0 real > 0 for t^n), n a positive integer ;

     (b) $t^n J_0(t)$, $t^n \log t$, $t^n Y_0(t)$ are bounded in $0 \leqslant t \leqslant h$, but not in $t \geqslant 0$;

     (c) $t^n K_0(t)$ is bounded in $t \geqslant 0$;

     (d) $e^{-p_0 t} x^t$ is bounded in $t \geqslant 0$; if x real > 0, $p_0 \geqslant \log x$.

See captions to Figs. 2, 3.

2. Draw the graphs of and define the derivatives of the functions in Figs. 20, 25b.

3. Demonstrate that $f(t) = 1/t$ is bounded in the range $t \geqslant h$, $h > 0$ but *not* in $t > 0$.

4. Show that $\cos(1/x)$ has a finite oscillatory discontinuity at the origin.

5. Show that $\dfrac{1}{x} \cos(1/x)$ has an infinite oscillatory discontinuity at the origin.

6. What is the order of infinity of $J_{-1/2}(t)$ at the origin?   [1/2.]

7. What is the order of infinity of (a) $t^{-\nu} \sin t$, $1 < \nu < 2$; (b) $1/\sin t$ at the origin?   [(a) $\nu - 1$, (b) 1.]

8. Show that if $f(t)$ is monotonic and positive, $\displaystyle\int_0^\infty e^{-pt} f(t)\,dt$ represents a monotonic positive function in that range of p for which the integral is uniformly convergent.

9. Without evaluating the Laplace integral, show that the L.T. of $\sin t$ is monotonic in the range p real > 0 ; also show that $J_0(it)$ is a positive monotonic increasing function which $\rightarrow +\infty$ with t.

10. The L.T. of a monotonic positive function cannot be alternating. Show that although $H_0(t)$ and $Y_0(t)$ are both alternating, their difference $H_0(t) - Y_0(t)$ is a monotonic positive function in the range $t \geqslant 0$, unbounded at the origin.

11. Show that $\sin t \operatorname{ci}(t) - \cos t \operatorname{si}(t)$ is a monotonic positive function in the range $t > 0$. [See Problem 10.]

* Additional technical problems (with answers) soluble by the L.T. method will be found in reference 12.

184

12. Prove that the following series are uniformly convergent in any closed interval of t, $h_1 \leqslant t \leqslant h$:

(a) $J_0(t) = \sum\limits_{r=0}^{\infty} (-1)^r \dfrac{(\frac{1}{2}t)^{2r}}{(r!)^2}$; (b) $I_0(t) = \sum\limits_{r=0}^{\infty} (\frac{1}{2}t)^{2r}/(r!)^2$;

(c) $J_n(t) = \sum\limits_{r=0}^{\infty} (-1)^r \dfrac{(\frac{1}{2}t)^{n+2r}}{r!(n+r)!}$; (d) $I_n(t) = \sum\limits_{0}^{\infty} \dfrac{(\frac{1}{2}t)^{n+2r}}{r!(n+r)!}$;

(e) $\mathsf{H}_n(t) = \sum\limits_{r=0}^{\infty} (-1)^r \dfrac{(\frac{1}{2}t)^{n+2r+1}}{\Gamma(r+3/2)\,\Gamma(n+r+3/2)}$,

n a positive integer > 0.

13. Prove that each series in Problem 12 may be differentiated term by term over the interval $t = (0, h)$.

14. Prove that each series in Problem 12 may be integrated term by term over the interval $t = (0, h)$.

15. Show that the following series are absolutely and uniformly convergent in any closed interval of t or w, as the case may be :

(a) $2\sqrt{t/\pi} \left\{ 1 - \dfrac{z^2}{1.3.5} + \dfrac{z^4}{1.3.5.7.9} - \ldots \right\}$, $z = \omega t$, ω real > 0 ;

(b) $\dfrac{4w^{3/2}}{3\sqrt{\pi}} \left\{ 1 + \dfrac{v^2}{5.7.9} + \dfrac{v^6}{5.7.9\ldots 15} + \ldots \right\}$, $v = 2w$;

(c) $\sum\limits_{n=1}^{\infty} w^{3n}/(3n)!$

16. Demonstrate that the series $\sum\limits_{n=1}^{\infty} (-1)^n e^{-n^2 a^2 t}$, a real, is u.c. in any closed interval if $t > 0$, and oscillatory when $t = 0$. Is term by term differentiation with respect to t permissible? [Yes! if $t > 0$.]

17. Investigate the convergence of the series in (35), (42) § 8·4 ; (29), (30), (32), (33), (35)–(37) § 8·5 ; (2), (3) § 8·6.

18. Show that the series $\operatorname{erf} t = (2/\sqrt{\pi}) \left[t - \dfrac{t^3}{1!3} + \dfrac{t^5}{2!5} - \dfrac{t^7}{3!7} + \ldots \right]$ is absolutely and uniformly convergent in any closed interval of t.

19. Prove that term by term differentiation and integration of the series in Problem 18 is valid, and state the corresponding interval of t.

20. Show that the series $J_0(\sqrt{t^2 + 2bt}) = \sum\limits_{r=0}^{\infty} (-1)^r J_r(t) b^r/r!$ is absolutely and uniformly convergent in any finite intervals of t and b.

21. Prove that the following integrals may be differentiated under the integral sign with respect to p and b :

(a) $\displaystyle\int_0^{\infty} e^{-pt} J_0(bt)\,dt$; (b) $\displaystyle\int_0^{\infty} e^{-pt} \sinh bt\,dt$; (c) $\displaystyle\int_b^{\infty} e^{-pt} J_0(a\sqrt{t^2 - b^2})\,dt$.

State the six results obtained.

$$\Big[(a) \; -\int_0^\infty e^{-pt}t J_0(bt)\,dt = -p/(p^2+b^2)^{3/2},$$

$$-\int_0^\infty e^{-pt}t J_1(bt)\,dt = -b/(p^2+b^2)^{3/2};$$

(b) $\; -\int_0^\infty e^{-pt}t \sinh bt\,dt = -2bp/(p^2-b^2)^2,$

$$\int_0^\infty e^{-pt}t \cosh bt\,dt = (p^2+b^2)/(p^2-b^2)^2;$$

(c) $\; -\int_b^\infty e^{-pt}t J_0(a\sqrt{t^2-b^2})\,dt$

$$= -pe^{-b\sqrt{p^2+a^2}}(b+1)/\sqrt{p^2+a^2})/(p^2+a^2),$$

$$-\int_b^\infty e^{-pt}\frac{J_1(a\sqrt{t^2-b^2})\,dt}{\sqrt{t^2-b^2}} = (e^{-b\sqrt{p^2+a^2}} - e^{-bp})/ab \Big].$$

22. Prove that integral (1) § 6·12 may be differentiated under the integral sign with respect to p and ν.

23. Demonstrate the validity of differentiating $\int_0^\infty e^{-pt}\sin bt\,dt$ under the integral sign with respect to p. Thence show that

$$\int_0^\infty e^{-pt}t \sin bt\,dt = 2pb/(p^2+b^2)^2.$$

24. The asymptotic form of a certain function is

$$g(b,\,t) \sim \int_0^\infty e^{-ut}u \sin b\sqrt{u}\,du.$$

Show when t is large, b real > 0, that $g(b,\,t) \sim \dfrac{\sqrt{\pi}be^{-b^2/4t}}{8t^{5/2}}\Big(-\dfrac{b^2}{t}+6\Big).$

[Use the L.T. of $\sin b\sqrt{u}$, and differentiate the corresponding infinite integral under the sign with respect to p, i.e. use (3) § 2·17.]

8·2. Convergence and evaluation of integrals, inverting the order of integration.

1. Show that $\int_0^\infty e^{-pt}f(t)\,dt$ is absolutely and uniformly convergent in $p_1 \leqslant p \leqslant p_2$, p_1 real > 0, when $f(t) = \sin t$, $\cos t$, $\cos^3 t$, $t^n \sin^3 t$, $t^n \cos^3 t$, $\sin^n t$, $\cos^n t$, $t^n J_0(t)$, $J_0{}^n(t)$, $J_\nu(t)$, $J_\nu{}^n(t)$, $R(\nu)> -1$, $J_\nu{}^n(t)/t$, $R(\nu)>0$, $t^\nu \log t$, $t^\nu \log^n t$, $R(\nu)> -1$, $n \geqslant 1$, $J_0(t)/t^\nu$, $R(\nu)<1$.

2. In Problem 1 if $a\sqrt{t^2-b^2}$ is written for t in $f(t)$ and b for 0 in the lower limit, prove that all the integrals are absolutely and uniformly convergent.

3. Demonstrate that $\int_0^\infty e^{-pt}f(t)\,dt$ is uniformly convergent with respect to p in $p>1$ if $f(t)=\sinh t$, $\cosh t$, and in p real $> a$ if $f(t)=I_\nu(at)$, $R(\nu)>-1$.

4. Verify that

(a) $\int_0^\infty e^{-x^2/4t}\cos x\,dx$ is uniformly convergent with respect to t in the interval $h_1 \leqslant t \leqslant h$, $h_1>0$;

(b) $\int_0^\infty e^{it}\sin t\,dt$ is divergent ;

(c) $\int_0^\infty \sin \sqrt{t}\,\sin t\,dt$ is divergent ; plot $\sin \sqrt{t}\,\sin t$, $t=(0,\,8\pi)$;

(d) $\int_0^\infty \sin ay^n\,dy$ is u.c. with respect to a (real) in $h_1 \leqslant a \leqslant h$, $h_1>0$, $n\geqslant 2$. The integrand is alternating but not periodic, since the interval between consecutive zeros decreases with increase in y. Plot $\sin y^n$, $n=2, 4$.

(e) Using the footnote on p. 165, show that the value of the integral in (d) above is $\pi/2a^{1/n}\,n\Gamma(1-1/n)\cos(\pi/2n)$. [Write $ay^n=t$.]

(f) Show that the integral (2) § 5·151 is uniformly convergent with respect to t in $h_1 \leqslant t \leqslant h$, $h_1>0$. [See $5°$ § 5, Appendix III, respecting (c), (f).]

5. Prove that $\int_0^\infty e^{-pt}t^{\nu/2}J_\nu(i^{3/2}t^{1/2})\,dt$ converges uniformly with respect to p in the range p real >0, if $R(\nu)>-1$. The latter restriction is essential for convergence at the lower limit. [See § 1·21 and $2°$ § 8, Appendix III.]

6. Determine the conditions for convergence of $\int_0^\infty e^{-pt}J_\nu(at)\,dt$, p and a being complex. [$R(p+ia)>0$, $R(p-ia)>0$ at the upper limit, $R(\nu)>-1$ at the lower limit. See § 1·21 and $2°$ § 8, Appendix III. The first two conditions, which pertain to the upper limit, may be obtained by using the asymptotic form of the Bessel function. This includes a factor of the type $\cos[(x+iy)t-\alpha]$, expressible in terms of exponentials.]

7. Show that $\int_0^\infty g(t)\,dt$ is divergent when $g(t)=t^m J_n(t)$; $J_n^2(t)$; $t^m J_n^2(t)$; m integral $\geqslant 1$, n integral $\geqslant 0$; $|J_\nu(t)|$, $R(\nu)>-1$; $\sin \sqrt{t}$; $Y_0(\sqrt{t})$; $H_0(t)$ [use $H_0(t)\sim Y_0(t)+\dfrac{2}{\pi t}$ when t is large] ; $J_0(t)\sin t$; $J_0(t)\cos t$; $e^{\pm i\omega t}I_0(t)$; $\sin 1/t$; $J_0(1/t)$; [in two latter cases use lower limit 1]. Explain why each integral is divergent.

8. Evaluate $\int_{h_1}^{\infty} f(t)\,dt$ when $f(t) = (a)\ J_\nu(t),\ R(\nu) > -1$;

$(b)\ J_\nu(at),\ R(\nu) > -1$; $(c)\ J_\nu(t)/t,\ R(\nu) > 0$; $(d)\,t^{-1/2}\sin t$,

$(e)\ K_0(t)$; $(f)\ K_\nu(at),\ -1 < R(\nu) < 1$; $(g)\ tK_0(at)$; $(h)\ tK_1(t)$;

$(j)\ J_0(a\sqrt{t^2-b^2})$; $(k)\ J_1(a\sqrt{t^2-b^2})/\sqrt{t^2-b^2}$; $(l)\ tJ_1(a\sqrt{t^2-b^2})/\sqrt{t^2-b^2}$.

In (a)–(h) inclusive $h_1 = 0$, whilst in (j)–(l) $h_1 = b$.   [(a) 1; (b) $1/a$; (c) $1/\nu$; (d) $\sqrt{\pi/2}$; (e) $\frac{1}{2}\pi$; (f) $\pi/2a\cos\frac{1}{2}\nu\pi$; (g) $1/a^2$; (h) $\frac{1}{2}\pi$; (j) e^{-ab}/a; (k) $(1-e^{-ab})/ab$; (l) $1/a$.   Taking $a = 1$, $\nu = -1/2$ in (b) show that
$$C(t) = \frac{1}{\sqrt{2\pi}}\int_0^t \cos x\,dx / \sqrt{x} \to \tfrac{1}{2}, \text{ when } t \to +\infty. \text{ In } (b) \text{ if } \nu = 1/2,$$

$$S(t) = \frac{1}{\sqrt{2\pi}}\int_0^t \sin x\,dx / \sqrt{x} \to \tfrac{1}{2} \text{ when } t \to +\infty, \text{ so } \lim C(t) = \lim S(t).$$

9. Confirm the following results :

$(a)\ \displaystyle\int_0^{\infty} J_0(t) \left\{ \begin{matrix} \cos at \\ \sin at \end{matrix} \right\} dt = \left\{ \begin{matrix} 1/\sqrt{1-a^2}, \\ 0 \end{matrix} \right. \quad 0 \leqslant a < 1$;

$(b)\ \displaystyle\int_0^{\infty} e^{\pm iat} K_0(t)\,dt = (\tfrac{1}{2}\pi \pm i \sinh^{-1}a)/\sqrt{1+a^2},\ a \text{ real} \geqslant 0$;

$(c)\ \displaystyle\int_0^{\infty} K_0(t) \left\{ \begin{matrix} \cos at \\ \sin at \end{matrix} \right\} dt = \left\{ \begin{matrix} \pi/2\sqrt{1+a^2} \\ (\sinh^{-1}a)/\sqrt{1+a^2} \end{matrix} \right\} \begin{matrix} a \geqslant 0 \\ a > 0 \end{matrix}$;

$(d)\ \displaystyle\int_0^{2a} \sqrt{2at - t^2} \left\{ \begin{matrix} \cos \omega t \\ \sin \omega t \end{matrix} \right\} dt = (\pi a/\omega) J_1(a\omega) \left\{ \begin{matrix} \cos a\omega \\ \sin a\omega \end{matrix} \right\},\ a \text{ real} > 0$.

[Write $\pm ia$ or $\pm i\omega$ for p in the appropriate Laplace integral, prove that it is convergent, use the list, equate real and imaginary parts in (a) (c), (d).]

10. Confirm the following results :

$(a)\ \omega_c^{2n+1}\displaystyle\int_0^{\infty} \sin\omega t\,J_{2n}(\omega_c t)\,dt = \sin(2n\sin^{-1}a)/\sqrt{1-a^2},\ 0 < a < 1$;

$$= (-1)^n/\sqrt{a^2-1}\,(a + \sqrt{a^2-1})^{2n},\ a > 1$$;

$(b)\ \omega_c^{2n+1}\displaystyle\int_0^{\infty} \cos\omega t\,J_{2n}(\omega_c t)\,dt = \cos(2n\sin^{-1}a)/\sqrt{1-a^2},\ 0 < a < 1$;
$$= 0, \qquad\qquad a > 1$$;

(c) $\omega_c \displaystyle\int_0^\infty \mathbf{H}_0(\omega_c t) \begin{Bmatrix} \cos \omega t \\ \sin \omega t \end{Bmatrix} dt = \begin{cases} 1/\sqrt{1-a^2}, \\ 2 \cosh^{-1}(1/a)/\pi\sqrt{1-a^2} \end{cases}, 0 < a < 1,$

where $a = \omega/\omega_c$, n a positive integer.

(d) Show that

$\omega_c{}^{2n+1} \displaystyle\int_0^\infty e^{\pm i\omega t} J_{2n}(\omega_c t) dt = [T_{2n}(\cos \theta) \pm i U_{2n}(\cos \theta)]/\cos \theta,$

with $\cos \theta = \sqrt{1-a^2}$, $\omega_c > \omega$. T_{2n}, U_{2n} are Tchebychev's functions of the first and second kinds of order $2n$. If

$\qquad P = \{t + i(1 - t^2)^{1/2}\}^{2n}, Q = \{t - i(1 - t^2)^{1/2}\}^{2n},$

then $\qquad T_{2n} = \frac{1}{2}(P + Q) = \cos(2n \cos^{-1} t),$

and $\qquad U_{2n} = \dfrac{1}{2i}(P - Q) = \sin(2n \sin^{-1} t).$

[Prove that $\displaystyle\int_0^\infty e^{-i\omega t} f(\omega_c t) dt$ converges if

$\qquad \omega_c \neq \omega$ and $f(\omega_c t) = J_{2n}(\omega_c t)$, etc. ;

write $i\omega$ for p in formulae in the list.]

11. Justify the inverted order of integration in (a) § 5·161 ; (b) § 5·162.

12. Verify that

$\displaystyle\int_0^\infty e^{-t^2} dt \int_0^\infty e^{-u(1+p^2/4t^2)} u^{\frac{1}{2}(\nu-1)} du = \int_0^\infty e^{-u} u^{\frac{1}{2}(\nu-1)} du \int_0^\infty e^{-t^2-up^2/4t^2} dt.$

13. Evaluate the t integral on the r.h.s. of Problem 12, make a suitable change of symbols and obtain the result

$$p \int_0^\infty e^{-px-x^2} x^\nu dx = x^\nu e^{-x^2}.$$

14. Prove that

$$\int_0^\infty e^{-c^2 t} dt \int_0^\infty e^{-a^2 x^2 t} dx = \int_0^\infty dx \int_0^\infty e^{-(a^2 x^2 + c^2)t} dt = \pi/2ac.$$

15. Establish the validity of the inverted order of integration in the following case :

$$p \int_0^\infty e^{-px} \left[\int_0^\infty \frac{t \sin xt}{(1 + t^2)} dt \right] dx = p \int_0^\infty \frac{t \, dt}{(1 + t^2)} \int_0^\infty e^{-px} \sin xt \, dx,$$

and show that the value of the repeated integral is $\pi p/2(p+1)$.

16. Prove that

$$p \int_0^\infty e^{-pt} \left[\int_0^\infty \frac{\sin xt}{x(1 + x^2)} dx \right] dt$$

$$= p \int_0^\infty \frac{dx}{x(1 + x^2)} \int_0^\infty e^{-pt} \sin xt \, dt = \pi/2(p+1).$$

8·3. Derivation of theorems. Establish the theorems given below ; *specify all the essential conditions,* and enunciate each theorem formally as in Chapter II.

1. $a^t f(bt) = \left(\dfrac{p}{p - \log a}\right) \phi \left(\dfrac{p - \log a}{b}\right).$

2. $f(t - \log a) = a^{-p} \phi(p).$

3. $e^{-at} \displaystyle\int_0^t f(\tau)\, d\tau = \dfrac{p}{(p + a)^2} \phi(p + a).$

4. $(t\, d/dt)^n f(t) = (-1)^n \left(p \cdot \dfrac{d}{dp}\right)^n \phi(p),$ where

$$\left(v \frac{d}{dv}\right)^2 g(v) = v \frac{d}{dv}\left[v \frac{d}{dv} g(v)\right].$$

5.   (a) $t^n f'(t) = (-1)^n p \phi^{(n)}(p)$;

       (b) commencing with §2·13, show that $tf(t) = -p \dfrac{d}{dp}\!\left(\dfrac{\phi(p)}{p}\right).$

6.   (a) $t^{-n} f(t) = p \displaystyle\int_p^\infty \cdots \int_p^\infty \dfrac{\phi(p)}{p} (dp)^n$;

       (b) $\dfrac{f(t - a)}{t} = p \displaystyle\int_p^\infty e^{-ap} \phi(p) \dfrac{dp}{p}.$

7. $\left(\dfrac{1}{t} \dfrac{d}{dt}\right)^n f(t) = p \displaystyle\int_p^\infty p \int_p^\infty \cdots p \int_p^\infty \phi(p) (dp)^n,$

       if $\left[\left(\dfrac{1}{t} \dfrac{d}{dt}\right)^r f(t)\right]_{t=0} = 0,$ for $r = 0, 1, 2, \ldots n - 1.$

8.   (a) $\displaystyle\int_0^t t \int_0^t t \ldots \int_0^t t f(t) (dt)^n = (-1)^n p \left(\dfrac{1}{p} \dfrac{d}{dp}\right)^n [\phi(p)/p]$;

       (b) if $f_1(t) = \phi_1(p),\ f_2(t) = \phi_2(p),$

       then $\displaystyle\int_0^\infty f_1(t) \phi_2(t)\, dt/t = \int_0^\infty \phi_1(p) f_2(p)\, dp/p.$

[Commence with $\displaystyle\int_0^\infty e^{-pt} f_1(t)\, dt = \phi_1(p)/p,$ multiply both sides by $f_2(p)$ and integrate with respect to p from 0 to $+\infty$.]

9. $\displaystyle\int_0^t f(t - \lambda) f(\lambda)\, d\lambda = \phi^2(p)/p.$

10. $f(t) - a \displaystyle\int_0^t e^{-a\lambda} f(t - \lambda)\, d\lambda = \dfrac{p}{(p + a)} \phi(p).$

11. $\displaystyle\int_0^t J_0(\sqrt{t^2 - x^2}) f(x)\, dx = \dfrac{p}{p^2 + 1} \phi(\sqrt{p^2 + 1}).$ [Use §2·25.]

12. $\int_0^t I_0(\sqrt{t^2-x^2})f(x)\,dx = \dfrac{p}{p^2-1}\phi(\sqrt{p^2-1})$. [Use § 2·25.]

13. $\int_0^\infty \sqrt{t/x}\,J_1(2\sqrt{tx})f(x)\,dx = \phi(1/p)$. [Use § 2·25.]

14. $\int_0^\infty \dfrac{t^x f(x)}{\Gamma(x+1)}\,dx = \dfrac{\phi(\log p)}{\log p}$. [Use § 2·25 and take $\phi_1 = 1/\log p$, $h(p) = \log p$.]

15. Show that if $e^{-p\cdot t}f(t)$, $e^{-p\cdot t}f'(t)$ are bounded and continuous in $t \geqslant 0$, then $\int_0^\infty J_0(2\sqrt{tx})f'(x)\,dx = \phi(1/p) - f(0)$.

16. $t^{\frac12 \nu}\int_0^\infty x^{-\frac12 \nu}J_\nu(2\sqrt{xt})f(x)\,dx = p^{1-\nu}\phi(1/p)$, $R(\nu) > -1$. [Use §2·25.]

17. If $f_1(t) = \phi(p)$ and $f_2(t) = f_1(p)$, then
$$\int_0^\infty \frac{f_2(x)\,dx}{(p+x)^2} = \int_0^\infty e^{-pt}f_1(t)\,dt.$$

18. $\int_0^\infty \left[\int_0^t f_1(\lambda)f_2(t-\lambda)\,d\lambda\right]\dfrac{dt}{t} = \int_0^\infty \phi_1\phi_2\,dp/p^2$.

19. In § 2·24 take $f_1(t) = \phi_1(p)$, $f_2(t) = I(t) = p$, and we obtain
$$\int_0^t f_1(\lambda)I(t-\lambda)\,d\lambda = \int_0^t f_1(t-\lambda)I(\lambda)\,d\lambda = f_1(t).$$

Discuss the validity of this result.

20. If $f_1(t) = \phi_1(p)$, $f_2(t) = \phi_2(p)$, $f_3(t) = \phi_3(p)$,
$$\int_0^t \int_0^{\lambda_2} f_1(\lambda_1)f_2(\lambda_2)f_3(t-\lambda_1-\lambda_2)\,d\lambda_1\,d\lambda_2 = \phi_1\phi_2\phi_3/p^2, \quad t > \lambda_1 + \lambda_2.$$

21. Establish the theorems in 4–8 above if $f(t)$ is replaced by $f(a\sqrt{t^2-b^2})$.

22. If $f(a\sqrt{t^2-b^2}) = \phi(p)$,
$$\frac{d^n}{dt^n}f = p^n\phi(p) - e^{-bp}\sum_{r=0}^{n-1} p^{n-r}\left[\frac{d^r}{dt^r}f(a\sqrt{t^2-b^2})\right]_{t=b}.$$

23. $\int_b^t \dfrac{f(au)}{t}\,dt = \int_p^\infty \dfrac{\phi(p)}{p}\,dp$, $u = \sqrt{t^2-b^2}$, $t > b$.

24. $\int_t^\infty \dfrac{f(au)}{t}\,dt = \int_0^p \dfrac{\phi(p)}{p}\,dp$.

25. $\int_b^\infty f(au)\dfrac{dt}{t} = \int_0^\infty \phi(p)\dfrac{dp}{p}$.

26. Deduce (a) theorem (1) § 2·243 from § 2·25, (b) theorem 8b above by aid of § 2·25.

27. Show that

$$\int_t^\infty t\,f(t)\,dt = [\phi'(p) - \Phi(p)]/p - [\phi'(p) - \Phi(p)]/p \,_{p\to+0},$$

and state the requisite conditions.

28. If $f(x;\ \epsilon) = \phi(p;\ \epsilon)$
   suppose that $f(x;\ \epsilon) \to f(x)$ and $\phi(p;\ \epsilon) \to \phi(p)$ when $\epsilon \to 0$. Under what conditions does the relationship $f(x) = \phi(p)$ hold?

29. Commencing with $tf(t) = -p\dfrac{d}{dp}[\Phi(p)]$, apply §§ 2·21, 2·23 to show that $\displaystyle\int_0^t f(t)\,dt = \Big[\Phi(p)\Big]_\infty^p$, and that $\displaystyle\int_t^\infty f(t)\,dt = \Big[\Phi(p)\Big]_p^0$. State the conditions for the validity of these results.

30. Under what conditions is $\displaystyle\int_0^\infty f(t)\dfrac{dt}{t^2} = \int_0^\infty \phi(p)\,dp$? [Commence with $\displaystyle\int_0^\infty \phi(p)\,dp = \int_0^\infty \Big[p\int_0^\infty e^{-pt}f(t)\,dt\Big]\,dp$, and invert the order of integration.]

8·4. Laplace transforms, application of theorems.

1. Obtain the L.T.S. of the following functions :
   (a) $e^{-at}\sin^2 bt$;   (b) $e^{-at}\cos^2 bt$;   (c) $e^{-at}\sinh^2 bt$;
   (d) $e^{-at}\cosh^2 bt$;   (e) $e^{-at}\operatorname{erf} t$.

$$\Big[(a)\ 2b^2 p/(p+a)[(p+a)^2 + 4b^2]; \quad (b)\ \Big(\frac{p}{p+a}\Big)\Big[\frac{(p+a)^2 + 2b^2}{(p+a)^2 + 4b^2}\Big];$$

$$(c)\ 2b^2 p/(p+a)[(p+a)^2 - 4b^2]; \quad (d)\ \Big(\frac{p}{p+a}\Big)\Big[\frac{(p+a)^2 - 2b^2}{(p+a)^2 - 4b^2}\Big].$$

$$(e)\ \Big(\frac{p}{p+a}\Big)e^{(p+a)^2/4}\operatorname{erfc}\tfrac{1}{2}(p+a).\Big]$$

2. Use L.T.S. to verify the following recurrence formulae :
   (a) $J_0'(t) = -J_1(t)$;   (b) $2J_\nu'(t) = J_{\nu-1}(t) - J_{\nu+1}(t)$;
   (c) $tJ_\nu'(t) = \nu J_\nu(t) - tJ_{\nu+1}(t)$;   (d) $tI_\nu'(t) = tI_{\nu-1}(t) - \nu I_\nu(t)$;
   (e) $2\nu I_\nu(t) = t[I_{\nu-1}(t) - I_{\nu+1}(t)]$.

3. Using the recurrence formula $K_0'(t) = -K_1(t)$, show that $K_1(t)$ has no L.T., i.e. prove that $\displaystyle\int_0^\infty e^{-pt}K_1(t)\,dt$ is divergent.

4. Given that $I_0(\sqrt{t^2 - b^2}) = pe^{-b\sqrt{p^2-1}}/\sqrt{p^2 - 1}$, find the L.T. of $I_0(\sqrt{t^2 - 2bt})$. $[pe^{-b(p+\sqrt{p^2-1})}/\sqrt{p^2 - 1}]$, $t > 2b$.

5. Show that

$$\frac{d}{dt} I_0(a\sqrt{t^2 - b^2}) = at \frac{I_1(a\sqrt{t^2 - b^2})}{\sqrt{t^2 - b^2}} = p^2 \frac{e^{-b\sqrt{p^2 - a^2}}}{\sqrt{p^2 - a^2}} - pe^{-bp}.$$

[Use the L.T. of $I_0(a\sqrt{t^2 - b^2})$ and apply 22, § 8·3.]

6. Prove that

$$a^2 b^2 \int_b^t \frac{J_2(au)}{u^2} dt + a \int_b^t \frac{J_1(au)}{u} dt - \tfrac{1}{2} a^2 bH(t - b)$$
$$= \sqrt{p^2 + a^2} e^{-b\sqrt{p^2 + a^2}} - pe^{-bp},$$

where $u = \sqrt{t^2 - b^2}$. $\Big[$ Commence with

$$ab \int_b^\infty e^{-pt} \frac{J_1(au)}{u} dt = (e^{-bp} - e^{-b\sqrt{p^2 + a^2}}),$$

differentiate both sides with respect to b after proving the validity of this procedure, and use the recurrence formula

$$J_1'(u) = \frac{J_1(u)}{u} - J_2(u) \Big].$$

7. Invert $\dfrac{pe^{-bs}}{s} [1 + (p/a) - (s/a)]$, where $s = \sqrt{p^2 - a^2}$.

$$\Big[I_0(au) + (t - b) \frac{I_1(au)}{u} : \text{ use Example 5} \Big].$$

8. Show that if $f(t)$ is a monotonic increasing function and $\rightarrow A$, a constant, as $t \rightarrow + \infty$, the area between the curve $y = f(t)$ and its asymptote in the range $t = (0, +\infty)$ is $\lim_{p \to +0} [A - \phi(p)]/p$, provided $\int_0^\infty e^{-pt} [A - f(t)] dt$ converges uniformly as $p \rightarrow +0$.

9. In Problem 8, if $\phi(0) = A$, show that the area is given by $-\phi'(0)$, i.e. $\lim_{p \to +0} [A - \phi(p)]/p = -\phi'(0)$. [Expand using Maclaurin's theorem].

10. In Problem 8, if t is replaced by $a\sqrt{t^2 - b^2}$, the area over the range $t = (b, +\infty)$ is expressed by $\lim_{p \to +0} [Ae^{-bp} - \phi(p)]/p$.

11. Prove that the centre of gravity of the area in Problem 8 is distant $\lim_{p \to +0} [(1/p) + \phi'/(A - \phi)]$ from the axis of $f(t)$. If $\phi(0) = A$, show that the distance in question is $-\phi''(0)/2\phi'(0)$.

12. Prove that k^2, the square of the radius of gyration of the area in Problem 8 about the axis of $f(t)$, is

$$\lim_{p \to +0} \{(2/p^2) + [(2\phi'/p) - \phi'']/(A - \phi)\}.$$

13. In Problem 12, if $\phi(0) = A$, then $k^2 = \phi'''(0)/3\phi'(0)$.

14. Verify that $\int_0^t J_0[2\sqrt{x(t-x)}]\,x\,dx = \tfrac{1}{2}t\sin t$.

15. Show that $\int_0^t J_0(\sqrt{t^2-x^2})I(x)\,dx = J_0(t)$, $I(x)$ being an impulsive function as in § 2·19.

16. Show that

(a) $\int_0^t J_0(\sqrt{t^2-x^2})\,x\,dx = tJ_1(t)$;

(b) if $\int_0^\infty pxJ_0(x)\,dx/\sqrt{p^2x^2+1} = e^{-1/p}$,

then $\int_0^\infty J_0(x)J_0(t/x)\,dx = J_0(2\sqrt t)$, $t>0,\ p$ real >0.

17. Show that $\int_0^\infty e^{-pt}g(t)I(t)\,dt = 1$, if $g(0)=1$.

18. Show that $\int_0^t J_0(\sqrt{t^2-x^2})I(x-b)\,dx = J_0(\sqrt{t^2-b^2})$, $0<b<t$.

19. Show that (a) $\int_0^\infty e^{-x^2/4t}I(x-h)\,dx = e^{-h^2/4t}$, $h>0,\ t>0$;

(b) $\int_0^t I_0[2\sqrt{x(t-x)}]x\,dx = \tfrac{1}{2}t\sinh t$.   [Apply § 2·251].

20. If the area between $\omega\cos\omega t$ and the t axis from 0 to $\pi/2\omega$ is constant, show that as $\omega\to+\infty$, the L.T. $\to p$.

21. Show that

(a) $\int_0^\infty e^{-x^2/4t}x^{1/2}\,dx\int_0^\infty J_1(2\sqrt{xy})\cos y\,y^{-1/2}\,dy = \sqrt{\pi t}\,(1-e^{-t})$.

[See 16, § 8·3 and (3) § 2·253.]

(b) Using the formula in 20, § 8·3, establish the relationship

$\int_0^1\int_0^1 x^{\sigma-1}y^{\mu-1}(1-x-y)^{\nu-1}\,dx\,dy = \Gamma(\sigma)\Gamma(\mu)\Gamma(\nu)/\Gamma(\sigma+\mu+\nu)$,
$R(\sigma),\ R(\mu),$ and $R(\nu)>0$.

22. If in Problem 21 (a) $J_0(y)$ replaces $\cos y$, show that the result is $\sqrt{\pi t}\,\mathrm{erf}\,\sqrt t$.

23. Verify that $\int_0^\infty e^{-x^2/4t}\cosh x\,dx = \sqrt{\pi t}\,e^t$, $t>0$.

24. Show that if $(m-n)>-2$,

$\int_0^\infty e^{-x^2/4t}He_n(x/\sqrt{2t})x^m\,dx = 2^{\frac{1}{2}n}\sqrt\pi\,m!\,t^{\frac{1}{2}(m+1)}/\Gamma[\tfrac{1}{2}(m-n)+1]$.

25. Prove that $\int_0^\infty e^{-x^2/4t}\left[\int_0^x \operatorname{bei} y\,dy\right]dx = \sqrt{2\pi t}\,S(t)$.

26. Prove that $\int_0^\infty e^{-x^2/4t}\left[\int_0^x \operatorname{ber} y\,dy\right]dx = \sqrt{2\pi t}\,C(t)$.

27. Show that

$$\int_0^\infty e^{-u^2 t}J_0(au)J_0(bu)u\,du = e^{-(a^2+b^2)/4t}I_0(ab/2t), \qquad t>0.$$

28. Show that $\left[\int_0^t d\lambda\right]^n f(\lambda) = \int_0^t \frac{(t-\lambda)^{n-1}}{(n-1)!}f(\lambda)\,d\lambda, \quad n\geqslant 1.$ State any restrictions on $f(t)$.

29. Establish the following results and verify each step in so doing :

(a) $\int_0^\infty \dfrac{\cos xt}{(1+x^2)}\,dx = \tfrac{1}{2}\pi e^{-t}$;   (b) $\int_0^\infty e^{-tx^2}\,dx = \tfrac{1}{2}\sqrt{\pi/t}$;

(c) $\int_0^\infty \sin tx^2\,dx/x = \tfrac{1}{4}\pi$;   (d) $\int_0^\infty \dfrac{\sin xt}{x(1+x^2)}\,dx = \tfrac{1}{2}\pi(1-e^{-t})$;

(e) $\int_0^\infty \cos xt\,dx/\sqrt{x} = \sqrt{\pi/2t}$ given that $\int_0^\infty dy/(1+y^4) = \pi/2\sqrt{2}$.

30. Verify the following results :

(a) $\displaystyle\int_0^\infty e^{-t\sinh\theta}J_0(t)\,dt = \operatorname{sech}\theta$;

(b) $\displaystyle\int_0^\infty e^{-t\sinh\theta}Y_0(t)\,dt = -\frac{2\theta}{\pi}\operatorname{sech}\theta$;

(c) $\displaystyle\int_0^\infty e^{-t\sinh\theta}H_0^{(1),(2)}(t)\,dt = \left[1\mp\frac{2\theta i}{\pi}\right]\operatorname{sech}\theta$;

(d) $\displaystyle\int_0^\infty e^{-t\sinh\theta}J_\nu(t)\,dt = e^{-\nu\theta}\operatorname{sech}\theta, \qquad R(\nu)>-1$

(e) $\displaystyle\int_0^\infty e^{-t\sinh\theta}\sin t\,dt = \operatorname{sech}^2\theta$;

(f) $\displaystyle\int_0^\infty e^{-t\sinh\theta}\cos t\,dt = \tanh\theta\operatorname{sech}\theta,$ \qquad all for θ real >0.

[Put $\sinh\theta = p$ and use the list.]

31. Confirm the following results and justify the implied changes in the order of integration :

(a) $\displaystyle\int_0^t J_0(t-\lambda)J_0(\lambda)\,d\lambda = \sin t$;   (b) $\displaystyle\int_0^t J_0(t-\lambda)\sin\lambda\,d\lambda = tJ_1(t)$;

(c) $\displaystyle\int_0^t J_0(t-\lambda)J_\nu(\lambda)\,d\lambda/\lambda = J_\nu(t)/\nu, \qquad R(\nu)>0$;

(d) $\int_0^t I_0(\lambda)I_1(t-\lambda)\,d\lambda = \cosh t - I_0(t)$;

(e) $\int_0^t I_0(t-\lambda)I_0(\lambda)\,d\lambda = \sinh t$;

(f) $\int_0^t S(\lambda)C(t-\lambda)\,d\lambda = \tfrac14(t-\sin t)$;

(g) $\int_0^t I_0(t-\lambda)\sinh\lambda\,d\lambda = tI_1(t)$;

(h) $\int_0^t I_0(t-\lambda)I_\nu(\lambda)\,d\lambda/\lambda = I_\nu(t)/\nu,$ $R(\nu)>0$;

(j) $\int_0^t \sinh\lambda \sinh(t-\lambda)\,d\lambda = \tfrac12\sqrt{\pi/2}\int_0^t t^{3/2}I_{\frac12}(t)\,dt = \sqrt{\pi/8}\,t^{3/2}I_{3/2}(t)$;

(k) $\int_0^t J_{-\nu}(\lambda)J_\nu(t-\lambda)\,d\lambda = \sin t,$ $-1<R(\nu)<1$;

(l) $\int_0^t J_\nu(\lambda)J_\mu(t-\lambda)\,d\lambda/\lambda = J_{\mu+\nu}(t)/\nu,$ $R(\mu)>-1,\;\; R(\nu)>0,$
$\mu+\nu\neq 0$;

(m) $\int_0^t \lambda^\mu J_\mu(\lambda)(t-\lambda)^\nu J_\nu(t-\lambda)\,d\lambda$

$= \Gamma(\mu+1/2)\,\Gamma(\nu+1/2)\,t^{\mu+\nu+1/2}J_{\mu+\nu+1/2}(t)\big/\sqrt{2\pi}\,\Gamma(\mu+\nu+1),$
$R(\mu)>-\tfrac12,\;R(\nu)>-\tfrac12,\;\mu+\nu+\tfrac12\neq 0;$

(n) $\int_0^t J_1(t-\lambda)Ji_0(\lambda)\,d\lambda = Ji_0(t)+\tfrac12\pi Y_0(t),$ where

$$Ji_0(t) = \int_t^\infty J_0(x)\,d.c/x.$$

In each case, except (a), (e), (j), the result can be represented by a second integral which should be written down by the reader. Since $J_0(0)=1$, $(\mu+\nu)\neq 0$ in (l).

32. Invert each term of (5) § 7·12 and obtain the Maclaurin expansion of $f(t)$. Apply (5), (6) § 2·24 to the integral term and obtain the remainder

$$R_n = \int_0^t f^{(n)}(t-\lambda)\lambda^{n-1}\,d\lambda/(n-1)!$$

$$= \int_0^t f^{(n)}(\lambda)(t-\lambda)^{n-1}\,d\lambda/(n-1)! \quad \left[\phi_1\phi_2/p = \frac{1}{p}\cdot\frac{1}{p^{n-1}}\cdot p\int_0^\infty, \text{ etc.}\right].$$

33. Commencing with $f(t-h)$ and proceeding as in § 7·12, obtain the Taylor's series with remainder

$$R_n = \int_0^t f^{(n)}(t-h-\lambda)^{n-1}\,d\lambda/(n-1)! = \int_0^t f^{(n)}(\lambda-h)(t-\lambda)^{n-1}\,d\lambda/(n-1)!$$

[*Phil. Mag.*, **26**, 695, 1938.]

34. Show that if m and n are positive integers, or if $(m+n)$ is a positive integer,

$$\int_0^t J_{2m}(\lambda)J_{2n}(t-\lambda)\,d\lambda = \int_0^t J_{2n}(\lambda)J_{2m}(t-\lambda)\,d\lambda$$
$$= (-1)^{m+n}[\sin t - 2\{J_1(t) - J_3(t) + \ldots + (-1)^{m+n+1}J_{2m+2n-1}(t)\}]$$

$\Big[$ See procedure in § 5·144 :

$$\phi_1\phi_2/p = p\,b^{2m+2n}/(p^2+1) = \frac{2p}{\sqrt{p^2+1}}\cdot\frac{b^{2m+2n+1}}{(1+b^2)} = (-1)^{m+n+1}\frac{p}{\sqrt{p^2+1}}$$

$$\times\left\{-\frac{2b}{(1+b^2)} + \frac{2b}{(1+b^2)}[1+(-1)^{m+n+1}b^{2m+2n}]\right\};\ b = (\sqrt{p^2+1}-p)\Big].$$

35. (a) Taking

$$p/(p^2+1) = (p/\sqrt{p^2+1})2b/(1+b^2),\ \text{with}\ b = \sqrt{p^2+1}-p,$$

show that $\sin t = 2\displaystyle\sum_{r=0}^{\infty}(-1)^r J_{2r+1}(t)$.

(b) Taking $2pb = (1-b^2)$, show that

$$\cos t = J_0(t) + 2\sum_{r=1}^{\infty}(-1)^r J_{2r}(t).$$

36. Verify the following results by aid of § 2·22 :

(a) $\displaystyle\int_0^{\infty}[J_0(t) - \cos t]\,dt/t = \log_e 2$;

(b) $\displaystyle\int_0^{\infty} e^{-at}\sin bt\,dt/t = \tan^{-1}b/a$, $\qquad a \geqslant 0,\ b > 0$;

(c) $\displaystyle\int_0^{\infty}(e^{-t} - e^{-xt})\,dt/t = \log_e x$, $\qquad x\ \text{real} > 0$;

(d) $\displaystyle\int_0^{\infty}[e^{-ax} - J_0(bx)]\,dx/x = \log_e(b/2a)$, $\qquad a > 0,\ b > 0$;

(e) $\displaystyle\int_0^{\infty}[J_0(bx) - J_0(ax)]\,dx/x = \log_e(a/b)$, $\qquad a,\ b > 0$;

(f) $\displaystyle\int_0^{\infty}(\cos xt_1 - \cos xt_2)\,dx/x = \log_e(t_2/t_1)$, $\qquad t_1,\ t_2 \neq 0$.

Note that none of the integrands has an infinity at the origin, a point readily verified by expansion.

37. Verify that

$$\int_0^t J_0(2\sqrt{t-\lambda})\,d\lambda/\sqrt{\lambda} = e^{-1}\sum_{n=0}^{\infty}(-1)^n 2^{n+1/2} He_{2n+1}(\sqrt{2t})/(2n+1)!$$

38. Show that $\int_0^t L_n(t-\lambda)\,d\lambda/\sqrt{\lambda} = He_{2n+1}(\sqrt{t})$.

39. Show that (a) $e\cos 2t = \sum_{n=0}^{\infty}(-1)^n 2^n He_{2n}(t\sqrt{2})/(2n)!$

(b) $e^{-1}\cosh 2t = \sum_{n=0}^{\infty} 2^n He_{2n}(t\sqrt{2})/(2n)!$

Compare 42(a).

40. Show that

(a) $e\sin 2t = \sum_{n=0}^{\infty}(-1)^n 2^{n+1/2} He_{2n+1}(t\sqrt{2})/(2n+1)!$

(b) $e^{-1}\sinh 2t = \sum_{n=0}^{\infty} 2^{n+1/2} He_{2n+1}(t\sqrt{2})/(2n+1)!$

41. Consider the convergence of the four series in 39, 40, and verify that the conditions in § 2·18 are satisfied.

42. Confirm that (a) $\cosh(2\sqrt{t}) = e^{\frac{1}{2}t+1}\sum_{n=0}^{\infty}\dfrac{2^n}{(2n)!} D_{2n}(\sqrt{2t})$;

(b) $\dfrac{e^{-t}}{\sqrt{\pi t}} + \operatorname{erf}\sqrt{t} = \dfrac{2}{\sqrt{\pi t}}\sum_{n=0}^{\infty}\dfrac{2^n n!}{(2n)!} D_{2n}(2\sqrt{t})$;

(c) $\dfrac{\sqrt{\pi}}{2}\operatorname{erf}\sqrt{t} = \sum_{n=0}^{\infty}\dfrac{2^n n!}{(2n+1)!} D_{2n+1}(2\sqrt{t})$.

43. Obtain the expansions of the following functions from their L.T.S. :

(a) $J_1(t)$; (b) $I_1(t)$; (c) $\sin^2 t$; (d) $\cosh^2 t$;
(e) $\operatorname{ber}^2 2\sqrt{t} + \operatorname{bei}^2 2\sqrt{t}$; (f) $\operatorname{erf} 1/2\sqrt{t}$.

44. Given that $K_0(2\sqrt{t}) - \log\sqrt{t}I_0(2\sqrt{t}) = e^{1/p}\log p$, show that

$$K_0(2\sqrt{t}) = -[\gamma + \log\sqrt{t}I_0(2\sqrt{t})] + \sum_{r=1}^{\infty}\frac{t^r}{(r!)^2}\Omega(r+1),$$

where $\Omega(r+1) = 1 + \dfrac{1}{2} + \dfrac{1}{3} + \dots + \dfrac{1}{r} = \psi(r+1) + \gamma$.

45. Given that

$$J_\nu(t) = \sum_{r=0}^{\infty}(-1)^r(\tfrac{1}{2}t)^{\nu+2r}/r!\,\Gamma(\nu+r+1)$$
$$= p/\sqrt{p^2+1}\,(p+\sqrt{p^2+1})^\nu, \quad R(\nu) > -1,$$

by differentiating with respect to ν show that

$$\frac{2}{\pi}\left[\log(\tfrac{1}{2}t)J_\nu(t) - \sum_{r=0}^{\infty}\frac{(-1)^r(\tfrac{1}{2}t)^{\nu+2r}\psi(\nu+r+1)}{r!\,\Gamma(\nu+r+1)}\right]$$
$$= -2p\log(p+\sqrt{p^2+1})/\pi\sqrt{p^2+1}\,(p+\sqrt{p^2+1})^\nu, \quad R(\nu)\geqslant 0.$$

46. Using the relationship $\dfrac{2}{\pi}\left[\dfrac{\partial}{\partial \nu} J_\nu(t)\right]_{\nu=0} = Y_0(t)$, obtain the result

$$Y_0(t) = \frac{2}{\pi}\left[\log\left(\tfrac{1}{2}t\right)J_0(t) - \sum_{r=0}^{\infty}(-1)^r\frac{(\tfrac{1}{2}t)^{2r}\psi(r+1)}{(r!)^2}\right]$$

$$= -2p\log\left(p + \sqrt{p^2+1}\right)/\pi\sqrt{p^2+1}.$$

47. In 46 when $r=0$, the first term in the expansion is $-\psi(1)=\gamma$ Euler's constant. Show that

$$Y_0(t) = \frac{2}{\pi}\left[\{\gamma + \log\left(\tfrac{1}{2}t\right)\}J_0(t) - \sum_{r=1}^{\infty}(-1)^r\frac{(\tfrac{1}{2}t)^{2r}\Omega(r+1)}{(r!)^2}\right].$$

This should be compared with the formula in 44 when t is written for $2\sqrt{t}$ on both sides thereof. Also obtain the expansion for $K_0(t)$ using the relation $\left[\dfrac{\partial}{\partial \nu} I_\nu(t)\right]_{\nu=0} = -K_0(t)$.

48. In 43–47 confirm that the conditions in § 2·18 for term by term inversion are satisfied.

49. Prove that if $f(t)$ has a Maclaurin expansion, $\phi(p)\to f(0)$ as $p\to +\infty$. Test this, using functions from the list. See below (5) § 7·12.

50. Explain why the lemma in 49 is inapplicable to functions like $\log t$, $Y_0(t)$, $K_0(t)$, but holds for t^ν, ν non-integral and > 0.

8·5. Integral, ordinary differential, and partial differential equations.

1. Solve the integral equations:

(a) $f(t) - a\displaystyle\int_0^t f(\tau)\,d\tau = 1$;   (b) $f(t) = \cosh t - \displaystyle\int_0^t f(\lambda)I_1(t-\lambda)\,d\lambda$

(c) $f(t) = f_1(t) + \displaystyle\int_0^t f(\lambda)\sin(t-\lambda)\,d\lambda$, where $f_1(t)$ is given;

(d) $f(t) = f_1(t) + \displaystyle\int_0^t f(\lambda)e^{-(t-\lambda)}\,d\lambda$;

(e) $f(t) = f_1(t) + \displaystyle\int_0^t f(\tau)\,d\tau$;   (f) $\displaystyle\int_0^t f(\lambda)f(t-\lambda)\,d\lambda = \sinh t$.

$\Bigg[$(a) e^{at};   (b) $I_0(t)$;   (c) $f_1(t) + \displaystyle\int_0^t\int_0^\tau f_1(\tau)\,(d\tau)^2$;

(d) $f_1(t) + \displaystyle\int_0^t f_1(\tau)\,d\tau$;   (e) $\dfrac{d}{dt}\displaystyle\int_0^t e^\lambda f_1(t-\lambda)\,d\lambda$;   (f) $I_0(t)$.

Substitute the L.T. of each side of the equation.$\Bigg]$

2. Solve for $y(t)$ the equations :

(a) $y + \omega^2 \int_0^t \int_0^\tau y(\tau)(d\tau)^2 = 1$; (b) $\dot{y} + a^2 \int_0^t y(\tau) d\tau = a$.

[(a) $\cos \omega t$; (b) $\sin at + y_0 \cos at$. Substitute the L.T. of each side.]

3. Solve : $2\phi(p) + p \int_p^\infty \phi(p) dp = 0$.

$\left[\phi = \dfrac{A\sqrt{\pi}}{2} p e^{p^2/4} \text{ erfc } (\tfrac{1}{2}p), \; A \text{ a constant. Invert in terms of } t,\right.$

$\left.\text{solve and find the L.T.}\right]$

(a) $\phi(p) + p \int_p^\infty \dfrac{\phi(p)}{p} dp = a^2$; (b) $\ddot{y} + a^3 \int_0^t y \, dt = 1$, if $y_0 = y_1 = 0$.

$\left[(a) \; \phi = a^2 \{1 - p e^p Ei(p)\} ;\right.$

$\left.(b) \; \dfrac{1}{3a^2} \left\{e^{-at} + e^{\frac{1}{2}at}\left(\sqrt{3} \sin \dfrac{\sqrt{3}}{2} at - \cos \dfrac{\sqrt{3}}{2} at\right)\right\}\right].$

5. Solve: $\phi + \omega^2 p \int_p^\infty \int_p^\infty \dfrac{\phi}{p} (dp)^2 = a^2$.

$[\phi = a^2\{1 - \omega p[\sin \omega p \; \text{ci}(\omega p) - \cos \omega p \; \text{si}(\omega p)]\}]$; use theorem $(6(a))$ § 8·3, invert in terms of t, solve and find the L.T.]

6. Solve the following ordinary D.E. :

(a) $\ddot{y} + 4\dot{y} + 3y = 0$; (b) $\ddot{y} + 4\dot{y} + 3y = e^{-t}$;

(c) $m\ddot{y} + r\dot{y} + sy = 0$; (d) $\ddot{y} + 2\dot{y} + y = \sin t$,

the initial conditions being $y = y_0$, $y' = y_1$ at $t = 0$. Verify that each solution satisfies the corresponding equation and the initial conditions.

7. Solve :

(a) $D^4 x - x = B$; (b) $D^4 x + 9D^3 x + 29D^2 x + 39Dx + 18x = B$,

subject to quiescence at $t = 0$, $D = d/dt$.

$\left[(a) \; x = \tfrac{1}{2} B (\cosh t + \cos t - 2) ;\right.$

$\left.(b) \; x = B \left\{\dfrac{1}{3 \cdot 3!} - \dfrac{e^{-t}}{4} + \dfrac{e^{-2t}}{2} - \dfrac{e^{-3t}}{6}\left(t + \dfrac{11}{6}\right)\right\}\right].$

8. If $x^{(n)} = x_n$ at $t = 0$, what is $\phi_3(p)$ the L.T. of the initial conditions function for the equation

$$(D^4 + 10D^3 + 35D^2 + 50D + 24) x = B ?$$

$[\phi_3 = x_0(p^4 + 10p^3 + 35p^2 + 50p) + x_1(p^3 + 10p^2 + 35p)$
$+ x_2(p^2 + 10p) + x_3 p.]$

9. Solve the equations in Problem 7 if B is replaced by $I(t)$ the impulsive function of unit strength.

$$\left[(a)\ \ 2x = \sinh t - \sin t; \qquad (b)\ \ x = \frac{e^{-t}}{4} - e^{-2t} + \frac{e^{-3t}}{2}\left(t + \frac{3}{2}\right)\right].$$

10. Solve (1) § 3·11 if e^{-bt} is replaced by $I(t)$. $\left[y = \dfrac{\sin at}{a}\right].$

11. Solve (1) §3·11 if e^{-bt} is replaced by $I(t)$, and the initial conditions are

$$y = y_0,\ y' = y_1\ \text{at}\ t = 0.\quad \left[y = \frac{\sin at}{a} + y_0 \cos at + y_1 \frac{\sin at}{a}\right].$$

12. Solve the problem in § 3·18 if the applied p.d. were $AI(t)$.

$$[I = (A/L)\cos \omega t - \omega Q_0 \sin \omega t,\ \omega^2 = 1/LC.$$
$$A \text{ has dimensions p.d.} \times \text{time}].$$

13. Solve $D^3 y + y = I(t)$, subject to quiescence initially.

$$[y = \tfrac{1}{3}\{e^{-t} + e^{\frac{1}{2}t}(\sqrt{3}\sin \sqrt{3}t/2 - \cos \sqrt{3}t/2)\}].$$

14. Solve $D^2 y - a^2 y = e^{-bt}$, with $y = y_0,\ y' = y_1$ at $t = 0,\ b^2 \neq a^2$.

$$y = 1/(b^2 - a^2)\left\{e^{-bt} + \frac{b}{a}\sinh at - \cosh at\right\} + y_0 \cosh at + \frac{y_1}{a}\sinh at\Big].$$

15. Solve 14 if $b^2 = a^2$. $\left[y = \left(\dfrac{y_1}{a} + \dfrac{1}{2a^2}\right)\sinh at + y_0 \cosh at - te^{-at}/2a\right].$

16. A p.d. E is applied at $t = 0$ to an inductance L having resistance R, initially quiescent. If $E = E_0$ a constant, show that the current at any time $t > 0$ is $I = (E_0/R)(1 - e^{-Rt/L})$. If $E = E_0 f(t)$, where $f(t) = \phi(p)$, show that $I = (E_0/L)\displaystyle\int_0^t e^{-R\lambda/L}f(t - \lambda)\,d\lambda.$

17. A p.d. E is applied at $t = 0$ to a capacitance C in series with a resistance R, the circuit being quiescent initially. What is the current when $t > 0$, if $E = E_0$ a constant? $[I = (E_0/R)e^{-t/CR}.$ At $t = 0$ the uncharged capacitance behaves as a short circuit, and the current rises precipitately to the value E_0/R. Thereafter the capacitance is charged according to the law $e^{-t/CR}]$.

18. In Problem 17 if $E = f(t)$ show that

$$RI = f(t) - \frac{1}{CR}\int_0^t e^{-\lambda/CR}f(t - \lambda)\,d\lambda.$$

19. Confirm that a capacitance C in series with inductance L, each being shunted by resistance R, behaves as a pure resistance whatever the form of applied p.d., provided $CR^2 = L$.

20. An inductance L of resistance R_1 is connected in parallel with a capacitance C in series with a resistance R_2. Show that the combination behaves as a pure resistance to all forms of applied p.d., provided $R_1 = R_2 = R$, and $CR^2 = L$.

21. Investigate the case in Problem 19, if the inductance has inherent resistance R_1.

22. Solve the simultaneous linear D.E. :

$$\frac{d^2\chi}{dt^2} + \frac{d^2\xi}{dt^2} = Ae^{-at}\sin \omega t, \quad \frac{d\chi}{dt} - b\xi = 0, \text{ subject to quiescence initially.}$$

$$\left[\xi = A\omega\left\{\frac{1}{b(a^2+\omega^2)} - \frac{e^{-bt}}{b\{(a-b)^2+\omega^2\}} + \frac{e^{-at}\sin(\omega t + \theta_1)}{\omega\sqrt{a^2+\omega^2}\sqrt{(a-b)^2+\omega^2}}\right\},\right.$$

$$\theta_1 = \tan^{-1}\omega(2a-b)/(a^2-ab+\omega^2);$$

$$\chi = A\omega\left\{\frac{t}{(a^2+\omega^2)} - \frac{(1-e^{-bt})}{b\{(a-b)^2+\omega^2\}} + \right.$$

$$\left.+ \frac{b[\sin(\theta_1+\theta_2) - e^{-at}\sin(\omega t+\theta_1+\theta_2)]}{\omega(a^2+\omega^2)\sqrt{(a-b)^2+\omega^2}}\right\}, \quad \theta_2 = \tan^{-1}\omega/a\right].$$

23. A coil having inductance L is coupled electromagnetically to a similar coil shunted by a capacitance C, the mutual inductance being M. The system is quiescent, and at $t = 0$ a constant p.d. E_0 is applied to the first coil. What is the current I_2 through C when $t > 0$?

$$\left[I_2 = -\frac{E_0}{L}\left(\frac{k}{1-k^2}\right)\sin \omega t,\right.$$

where $k = M/L$, the coefficient of coupling, and $\omega = 1/(\sqrt{LC}\sqrt{1-k^2})\right]$.

24a. Solve the problem in §3·211 if $R_1 = R_2$, $L_1 = L_2$, $k = 1$.

$$\left[I_2 = -\frac{E_0}{2R}e^{-Rt/2L}\right].$$

24b. Solve §3·211 under the conditions in Problem 24a, but with $E = E_0\sin \omega t$.

$$\left[-(E_0/2R)\left\{e^{-\beta t}\beta^2/(\beta^2+\omega^2)\right.\right.$$

$$\left.\left.+ \frac{\omega}{\sqrt{\beta^2+\omega^2}}\cos(\omega t - \tan^{-1}(\beta/\omega))\right\}, \text{ with } \beta = R/2L\right].$$

25. The equations for an electron moving in a plane parallel to that of the paper, between two large flat parallel plates perpendicular thereto and separated by a distance h are :

$$m\frac{d^2z}{dt^2} = Ee - H\frac{e}{c}\frac{dx}{dt}; \quad m\frac{d^2x}{dt^2} = H\frac{e}{c}\frac{dz}{dt};$$

where m = mass of electron, e = charge of electron, E = electric field strength between plates in z direction, H = magnetic field strength in y direction, c = velocity of light, x, z horizontal and vertical co-ordinates in plane of paper. Take quiescent initial conditions, assume the electron to start from one of the plates and determine its trajectory when constant E and H are applied at $t = 0$.

$[x = A\,(\omega t - \sin \omega t), \quad z = A\,(1 - \cos \omega t), \quad A = Ec/\omega H, \quad \omega = He/mc$.

x, z are the co-ordinates of a cycloid generated by a point on a circle of radius $Ec/\omega H$ rolling on the plate with angular velocity ω. If $h > 2Ec^2m/H^2e$ the electron fails to reach the other plate. md^2z/dt^2, md^2x/dt^2 are the accelerational forces in the z and x directions; Ee is the force in the z direction due to the electric field, whilst $H\dfrac{e}{c}\dfrac{dz}{dt}$, $H\dfrac{e}{c}\dfrac{dx}{dt}$ are the forces in the z, x directions due to the respective currents $\dfrac{e}{c}\dfrac{dz}{dt}, \dfrac{e}{c}\dfrac{dx}{dt}$ in virtue of the motion of the electron.]

26. A heavy disk, having moment of inertia I about its axis, is suspended centrally by a vertical wire rigidly fixed at its upper end. The inertia of the wire may be neglected in comparison with its torque stiffness τ per radian. The system is quiescent, and at $t = 0$ a torque $T = T_0(1 - e^{-at})$ is applied which causes the disk to turn. What is the angular velocity and displacement at any time $t > 0$?

$$\left[\dot\theta = \frac{aT_0}{I\sqrt{a^2+b^2}}\left\{\frac{e^{-at}}{\sqrt{a^2+b^2}} + \frac{1}{b}\sin\,(bt - \tan^{-1}b/a)\right\};\right.$$

$$\theta = \frac{aT_0}{I\sqrt{a^2+b^2}}\left\{\frac{a}{b^2\sqrt{a^2+b^2}} + \frac{1 - e^{-at}}{a\sqrt{a^2+b^2}}\right.$$

$$\left.\left. - \frac{1}{b^2}\cos\,(bt - \tan^{-1}(b/a))\right\},\quad b = \sqrt{\tau/I}\,\right].$$

27. A *rigid* loud speaker diaphragm (conical shell) of effective mass m_e has an elastic centering device of axial stiffness s, and negligible inertia. The resistance due to frictional loss and sound radiation is r_e per unit axial velocity. The diaphragm is at rest in its equilibrium position, and at $t = 0$ a force $AI\,(t)$ is applied centrally and axially. What is the axial velocity when $t > 0$? Take $s > r_e^2/4m_e$ for an oscillatory system.

$$\left[v = \frac{Ae^{-at}}{\omega m_e}\,(\omega \cos \omega t - a \sin \omega t), \quad a = \tfrac{1}{2}\frac{r_e}{m_e}, \quad \omega = \sqrt{\frac{s}{m_e} - \frac{r_e^2}{4m_e^2}}\right].$$

At $t = 0$ the axial acceleration is infinite momentarily; at $t = +0$ the diaphragm commences to move with a velocity A/m_e. This is the 'forced' initial condition mentioned in Appendix VI. As $t \to +0$ the acceleration is $(dv/dt)_{t\to+0} = -Ar_e/m_e^2\,\Big]$.

28. An electrical relay has a resistance of 5 ohms and an inductance of 0·15 henry. It is connected in series with a resistance of

3 ohms and a battery of 100 volts. A current of 4 amperes through the winding actuates a contact which short-circuits the 3 ohm resistance. After connecting the battery, what time elapses before the current reaches 0·95 of its final value?

29. Solve the linear partial D.E. $\dfrac{\partial^2 E}{\partial x^2} = \mathbf{CR}\,\dfrac{\partial E}{\partial t}$, subject to the initial condition $E(x, t) = 0$, $t \to -0$, $0 \leqslant x \leqslant l$, and the boundary conditions (a) $E(0, t) = 0$, $0 \leqslant t < \infty$, (b) $E(x, t) = E_0$ a constant, $x = l$, $t > 0$. Here E_x is the p.d. in an unloaded cable at a point distant x from the *far* end $(x = 0)$, which is connected to earth. A battery of voltage E_0 is applied at $t = 0$ between $x = l$ and earth.

$$\left[E_x = E_0 \left\{ (x/l) + 2 \sum_{n=1}^{\infty} (-1)^n e^{-n^2\pi^2 t/l^2 \mathbf{CR}} \frac{\sin(n\pi x/l)}{n\pi} \right\} \right].$$

30. The current at any point of the cable in Problem 29 is given by $I_x = \dfrac{1}{\mathbf{R}}\dfrac{\partial E}{\partial x}$. Prove that the series for E_x is absolutely and uniformly convergent if $t > 0$, and hence show that for $0 \leqslant x \leqslant l$,

$$I_x = (E_0/\mathbf{R}l) \left\{ 1 + 2 \sum_{n=1}^{\infty} (-1)^n e^{-n^2\pi^2 t/l^2 \mathbf{CR}} \cos(n\pi x/l) \right\}.$$

31. Verify that (1), (2) § 4·21 may be combined to give the P.D.E. $\dfrac{\partial^2 E}{\partial x^2} = \mathbf{LC}\,\dfrac{\partial^2 E}{\partial t^2} + (\mathbf{CR} + \mathbf{LG})\dfrac{\partial E}{\partial t} + \mathbf{GR}E$, and obtain the transform equation $\dfrac{d^2\varphi}{dx^2} - \lambda^2\varphi = 0$, with $\lambda^2 = (\mathbf{L}p + \mathbf{R})(\mathbf{C}p + \mathbf{G})$.

32. Apply (8) § 5·27 to 30 when $x = l$, and show that

$$I_l = E_0\sqrt{\mathbf{C}/\pi t\mathbf{R}} \left\{ 1 + 2 \sum_{n=1}^{\infty} e^{-n^2 l^2 \mathbf{CR}/t} \right\}.$$

[Write $\pi^2 t/l^2 \mathbf{CR}$ for t on both sides of (8) § 5·27.]

33. A spherical copper ball of radius $a = 5$ cm. is at a uniform temperature of $\theta_0 = 15°$ C. At $t = 0$ it is plunged into water at $\theta_1 = 35°$ C. Assuming the water to be in circulation and its temperature drop negligible, what is the temperature at any radius $r < a$ when $t > 0$? The D.E. for heat diffusion in a sphere is

$$\frac{\partial^2 \theta}{\partial r^2} + \frac{2}{r}\frac{\partial \theta}{\partial r} - \frac{1}{k}\frac{\partial \theta}{\partial t} = 0.$$

$$\left[\theta = (\theta_1 - \theta_0) \left\{ 1 + 2 \sum_{n=1}^{\infty} (-1)^n e^{-\pi^2 n^2 kt/a^2} \frac{\sin(\pi n r/a)}{(\pi n r/a)} \right\} + \theta_0 \right].$$

34. Plot a curve showing the relationship between central temperature and time in Problem 33, if the diffusivity

$$k = 1\cdot13 \text{ cm.}^2 \text{ sec.}^{-1}.$$

35. Solve $\dfrac{\partial^2\theta}{\partial x^2} - c\,\dfrac{\partial\theta}{\partial t} = 0$, subject to the initial condition

$$\theta(x, t) = 20^\circ\,\text{C.},\quad t \to -0,\quad 0 \leqslant x \leqslant l,$$

and the boundary conditions

(a) $\theta(x, t) = -10^\circ\,\text{C.},\ x = 0,\ t > 0$; (b) $\theta(x, t) = 20^\circ\,\text{C.},\ x = l,\ t > 0$.

$$\left[\theta = 20^\circ\text{-}30^\circ\left\{\frac{l-x}{l}\cdot + 2\sum_{n=1}^{\infty}(-1)^n e^{-n^2\pi^2 t/cl^2}\sin[n\pi(l-x)/l]/n\pi\right\}\right].$$

36. In Problem 35, $\theta(x, t)$ is the temperature at time t of any point, in a homogeneous rectangular slab,* distant $(l - x)$ from the plane face at 20° C. The parallel face is at -10° C. as in the case of a refrigerator. To maintain the lower temperature, heat must be extracted at a rate $q = \dfrac{1}{R}\dfrac{\partial\theta}{\partial x}$ per unit area, R being the thermal resistance per unit area per unit length of slab in the direction of heat flow. Show that the series in Problem 35 is absolutely and uniformly convergent if $t > 0$, and that when $x = 0$, $t > 0$,

$$q = \frac{30}{Rl} + \frac{60}{Rl}\sum_{n=1}^{\infty}e^{-n^2\pi^2 t/cl^2}.\qquad 1/c = k \text{ the diffusivity.}$$

37. The partial D.E. for radial heat diffusion in a long circular cylinder of radius a and length l $(l \gg a)$ is $\dfrac{\partial^2\theta}{\partial r^2} + \dfrac{1}{r}\dfrac{\partial\theta}{\partial r} - \dfrac{1}{k}\dfrac{\partial\theta}{\partial t} = 0$, where $\theta(r, t)$ is the temperature at radius r and time t, k being the diffusivity. Solve this equation subject to the initial condition

$$\theta(r, t) = 0,\quad t \to -0,\quad 0 \leqslant r \leqslant a,$$

and the boundary conditions

(a) $\theta(r, t) = \theta_0$ a constant at $r = a$, $t > 0$;

(b) $\theta(r, t)$ finite at $r = 0$, $t > 0$.

$$\left[\theta = \theta_0\left\{1 - 2\sum_{n=1}^{\infty}e^{-\alpha_n^2 kt/a^2}\frac{J_0(\alpha_n r/a)}{\alpha_n J_1(\alpha_n)}\right\},\right.$$

$$\left.\text{where } \alpha_n \text{ is the } n\text{th positive zero of } J_0(u)\right].$$

8·6. Discontinuous functions, Fourier expansions.

1. Prove that the L.T. of $\left.\begin{array}{l}f(t) = t^n\\ = 0\end{array}\right\}\begin{array}{l}0 < t < h\\ t < 0, t > h\end{array}$ is

$$\phi(p; 0, h) = -e^{-ph}[h^n + nh^{n-1}p^{-1} + n(n-1)h^{n-2}p^{-2} + \ldots + n!p^{-n}] + n!p^{-n}.$$

---

* The edges of the slab are insulated thermally, the heat loss therefrom being negligible. Alternatively, the slab is large enough for loss from the edges to be too small to affect the temperature distribution over a proportionately large central section.

2. Show that the L.T. of a semicircle of radius a, commencing at $t=0$ and repeated indefinitely, is $f(t) = \pi a e^{-ap} I_1(ap)/(1 - e^{-2ap})$. What is the corresponding Fourier expansion?

$$\left[f(t) = a \left\{ \tfrac{1}{4}\pi + \sum_{n=1}^{\infty} (-1)^n \frac{J_1(n\pi)}{n} \cos(n\pi t/a) \right\} \right].$$

This series converges uniformly as $t \to +0$, also $f(0) = 0$, so

$$\tfrac{1}{4}\pi = \sum_{n=1}^{\infty} (-1)^{n+1} \frac{J_1(n\pi)}{n} \bigg].$$

3. A voltage pulse of rectangular form, having unit amplitude, duration h_1, is repeated at interval $h \gg h_1$. Show that the corresponding Fourier expansion is

$$(h_1/h) + \frac{2}{\pi} \sum_{n=1}^{\infty} \frac{1}{n} \cos \frac{2\pi n}{h}\left(t - \frac{h_1}{2}\right) \sin \frac{n\pi h_1}{h}.$$

[The form is akin to that in Fig. 22 but the duration of the pulse is small compared with the interval between consecutive pulses].

4. A function representing a radio-telegraph wave-form consists of n cycles of $\sin \omega t$ commencing at $t=0$, followed by an equal quiescent interval. What is the L.T. if the function is repeated indefinitely? $[\omega p/(p^2 + \omega^2)(1 + e^{-2n\pi p/\omega})]$.

5. What is the Fourier expansion corresponding to the function in Problem 4?

6. Find the L.T. and the Fourier expansion corresponding to the following continuous wave-form. Show that the series is absolutely and uniformly convergent.

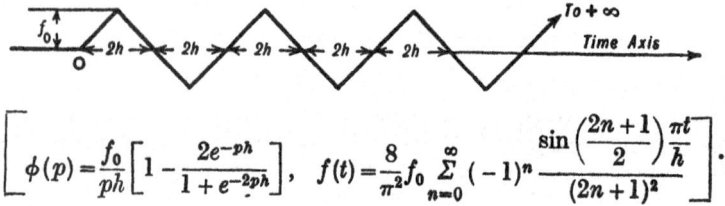

$$\left[\phi(p) = \frac{f_0}{ph}\left[\frac{1 - e^{-ph}}{1 + e^{-ph}}\right], \quad f(t) = \tfrac{1}{2}f_0\left[1 - \frac{8}{\pi^2} \sum_{n=0}^{\infty} \frac{\cos(2n+1)\pi t/h}{(2n+1)^2}\right]\right].$$

7. Derive the L.T. and the Fourier expansion for the following continuous wave-form. Show that the series is absolutely and uniformly convergent.

$$\left[\phi(p) = \frac{f_0}{ph}\left[1 - \frac{2e^{-ph}}{1 + e^{-2ph}}\right], \quad f(t) = \frac{8}{\pi^2}f_0 \sum_{n=0}^{\infty} (-1)^n \frac{\sin\left(\frac{2n+1}{2}\right)\frac{\pi t}{h}}{(2n+1)^2}\right].$$

8. Find the L.T. of the maritime distress signal ' SOS ' in Morse code as illustrated in the diagram. What is the L.T. if the signal is repeated indefinitely? S is represented by 3 dots, and O by three dashes, a dash being three times as long as a dot.

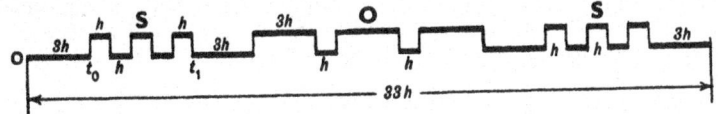

9. A note at the top of a pianoforte is repeated at equal time intervals h in a room acoustically dead. The sound lasts for a time $h_1 < h$ during each interval, the damper being effective for a time $(h - h_1)$. If overtones are negligible and the sound-wave has the form $e^{-\alpha t} \cos \omega t$ for an isolated note, what is the L.T. and the Fourier expansion corresponding to the repeated note?

10. A p.d. corresponding to the letter S in the Morse code (see Problem 8) is applied at $x = 0$ to a very long quiescent unloaded cable open at its far end. Find the p.d. and current at any point x, *after* each dot has been sent. The signal commences at t_0 and terminates at t_1, in the diagram of Problem 8, i.e. the p.d. $= 0$ for $t < t_0$ and $t > t_1$. The p.d. from $t = t_0$ to $(t_0 + h)$ is E_0. For convenience we take t_0 as 0.

$[E_1 = E_0 \{ \text{erfc} (K/\sqrt{t}) H(t) - \text{erfc} (K/\sqrt{t - h}) H(t - h) \}, \ h < t < 2h$;

$E_2 = E_1 + E_0 \{ \text{erfc} (K/\sqrt{t - 2h}) H(t - 2h) - \text{erfc} (K/\sqrt{t - 3h}) H(t - 3h) \},$

$$3h < t < 4h;$$

$E_3 = E_2 + E_0 \{ \text{erfc} (K/\sqrt{t - 4h}) H(t - 4h)$

$$- \text{erfc} (K/\sqrt{t - 5h}) H(t - 5h) \}, \ 5h < t < \infty, \ K = \tfrac{1}{2} x \sqrt{\mathbf{CR}}.$$

The formulae for the current are obtained on replacing $\text{erfc} (K/\sqrt{t - nh})$ by $\sqrt{[\mathbf{C}/\pi \mathbf{R} (t - nh)]} e^{-x^2 \mathbf{CR}/4(t - nh)}]$.

SHORT LIST OF
p-MULTIPLIED LAPLACE TRANSFORMS

The following abbreviations are used : $y = (t^2 - b^2)^{1/2}$; $r = (p^2 + a^2)^{1/2}$, $R = p + (p^2 + a^2)^{1/2}$; $s = (p^2 - a^2)^{1/2}$, $S = p + (p^2 - a^2)^{1/2}$. For simplicity we shall take a to be real. In many cases it may have either sign, e.g. in circular and exponential functions. t and b are real, $R(p) > 0$ usually.

1.
$$t^\nu = \Gamma(\nu + 1)/p^\nu \qquad\qquad R(\nu) > -1$$

2.
$$(t^2 - b^2)^\nu = (2b)^{\nu+1/2}\Gamma(\nu+1)K_{\nu+\frac{1}{2}}(bp)/\pi^{1/2}p^{\nu-1/2}, \; t > b,$$
$$R(\nu) \geqslant -\tfrac{1}{2}$$

3.
$$(1 + t^2)^{-1/2} = \tfrac{1}{2}\pi p[\mathbf{H}_0(p) - Y_0(p)]$$

4.
$$e^{-at} = p/(p+a)$$

5.
$$1 - e^{-at} = a/(p+a)$$

6.
$$e^{-at}t^\nu = \Gamma(\nu+1)p/(p+a)^{\nu+1} \qquad\qquad R(\nu) > -1$$

7.
$$e^{-a^2/4t}t^{-1/2} = (\pi p)^{1/2}e^{-ap^{1/2}}|$$

8.
$$\sin at = pa/(p^2 + a^2)$$

9.
$$(t/2a)\sin at = p^2/(p^2 + a^2)^2$$

10.
$$t^{-1}\sin at = p\tan^{-1}(a/p)$$

11.
$$t^{-1/2}\sin t = (\pi/2)^{1/2}p/(p^2+1)^{1/2}[p + (p^2+1)^{1/2}]^{1/2}$$

12.
$$\sin^2 t = 2/(p^2 + 4)$$

13.
$$t^{-1}\sin^2 t = \tfrac{1}{4}p\log(1 + 4/p^2)$$

14.
$$\cos at = p^2/(p^2 + a^2)$$

15.
$$\sin a\sqrt{t} = \tfrac{1}{2}a\sqrt{\pi/p}\,e^{-a^2/4p}$$

16.
$$e^{-at}\sin bt = pb/[(p+a)^2 + b^2]$$

17.
$$e^{-at}\cos bt = p(p+a)/[(p+a)^2 + b^2]$$

18.
$$\frac{1}{3a^2}\left[e^{-at} + e^{\frac{1}{2}at}\left(\sqrt{3}\sin\frac{\sqrt{3}}{2}at - \cos\frac{\sqrt{3}}{2}at\right)\right] = p/(p^3 + a^3)$$

19.
$$\sinh at = pa/(p^2 - a^2)$$

20.
$$\cosh at = p^2/(p^2 - a^2)$$

21.
$$\sinh^{-1} t = \tfrac{1}{2}\pi[\mathbf{H}_0(p) - Y_0(p)]$$

22.
$$\log t = -\gamma - \log p$$

23.
$$-Ei(-t) = \log(p+1) \qquad\qquad \text{(see § 5·21)}$$

24.
$$\mathrm{si}(t) = -\tan^{-1}p \qquad\qquad \text{(see § 2·231)}$$

25.
$$\mathrm{ci}(t) = -\tfrac{1}{2}\log(p^2 + 1) \qquad\qquad \text{(see § 2·231)}$$

26. $\qquad S(t) = [(p^2+1)^{1/2} - p]^{1/2}/2(p^2+1)^{1/2}$ (see 8, § 8·2)

27. $\qquad C(t) = [(p^2+1)^{1/2} + p]^{1/2}/2(p^2+1)^{1/2}$ (see 8, §8·2)

28. $\qquad \tan^{-1} t = \sin p \, \mathrm{ci}\,(p) - \cos p \, \mathrm{si}\,(p)$

29. $\qquad \mathrm{erf}\, t = e^{\frac{1}{4}p^2}(1 - \mathrm{erf}\, \tfrac{1}{2}p)$

30. $\qquad \mathrm{erf}\, t^{1/2} = 1/(p+1)^{1/2}$

31. $\qquad \mathrm{erfc}\,(a/2t^{1/2}) = e^{-ap^{1/2}}$

32. $\qquad a^{-1/2} t^{-3/2} x \, e^{-x^2/4at} = 2\sqrt{\pi p}\, e^{-x\sqrt{p/a}}.$

33. $\qquad e^t \, \mathrm{erf}\, t^{1/2} = p^{1/2}/(p-1)$

34. $\quad \tfrac{1}{2}\left[e^{-ab}\,\mathrm{erfc}\left(\dfrac{b-2at}{2t^{1/2}}\right) + e^{ab}\,\mathrm{erfc}\left(\dfrac{b+2at}{2t^{1/2}}\right)\right] = e^{-b(p+a^2)^{1/2}}$

35. $\quad \tfrac{1}{2}e^{a^2t}\left[e^{-ab}\,\mathrm{erfc}\left(\dfrac{b-2at}{2t^{1/2}}\right) + e^{ab}\,\mathrm{erfc}\left(\dfrac{b+2at}{2t^{1/2}}\right)\right] - \mathrm{erfc}\left(\dfrac{b}{2t^{1/2}}\right)$
$$= a^2 e^{-bp^{1/2}}/(p-a^2)$$

36. $\qquad J_0(at) = p/r \qquad\qquad\qquad\qquad$ (see 12a, § 8·1)

37. $\qquad J_\nu(at) = a^\nu p/rR^\nu \qquad$ (see 45, § 8·4) $\quad R(\nu) > -1$

38. $\qquad t^{-1}J_\nu(at) = a^\nu p/\nu R^\nu \qquad\qquad\qquad\qquad R(\nu) > 0$

39. $\qquad t^{\frac{1}{2}\nu}J_\nu[2(at)^{1/2}] = a^{\frac{1}{2}\nu} p^{-\nu} e^{-a/p} \qquad\qquad R(\nu) > -1$

40. $\qquad t^\mu J_\nu(2t^{1/2}) = \dfrac{\Gamma(\mu + \frac{1}{2}\nu + 1)}{\Gamma(\nu+1)p^{\mu+\frac{1}{2}\nu}}\, {}_1F_1(\mu + \tfrac{1}{2}\nu + 1;\, \nu+1;\, -1/p);$
$$R(\mu + \tfrac{1}{2}\nu) > -1$$

41. $\quad J_\nu(2at^{1/2}) J_\nu(2bt^{1/2}) = e^{-(a^2+b^2)/p} I_\nu(2ab/p) \qquad\quad R(\nu) > -1$

42. $\quad e^{-at}t^\nu J_\nu(bt) = (2b)^\nu \Gamma(\nu + \tfrac{1}{2}) p/[(p+a)^2 + 1]^{\nu+1/2}\pi^{1/2}$
$$R(\nu) > -\tfrac{1}{2}$$

43. $\quad e^{-at}t^{\frac{1}{2}\nu}J_\nu(2t^{1/2}) = pe^{-1/(p+a)}/(p+a)^{\nu+1} \qquad\qquad R(\nu) > -1$

44. $\qquad J_0(ay) = pe^{-br}/r \qquad\qquad\qquad\qquad\qquad\qquad\quad t > b$

45. $\qquad \displaystyle\int_b^t J_0(ay)\,dt = e^{-br}/r \qquad\qquad\qquad\qquad\qquad\quad t > b$

46. $\qquad \displaystyle\int_t^\infty J_0(ay)\,dt = \dfrac{e^{-b(p+a)}}{a} - \dfrac{e^{-br}}{r} \qquad\qquad\quad t > b$

47. $\qquad tJ_0(ay) = p^2 e^{-br}(b+1/r)/r^2 \qquad\qquad\qquad\qquad t > b$

48. $\qquad abJ_1(ay)/y = p(e^{-pb} - e^{-br}) \qquad\qquad\qquad\qquad\; t > b$

49. $\qquad tJ_1(ay)/y = (p/a)\left[e^{-pb} - \dfrac{p}{r}e^{-br}\right] \qquad\qquad\quad t > b$

50. $\qquad \left(\dfrac{t-b}{t+b}\right)^{\frac{1}{2}\nu} J_\nu(ay) = a^\nu pe^{-br}/rR^\nu \qquad\quad t > b,\, R(\nu) > -1$

51. $\qquad Y_0(at) = -2p \log (R/a)/\pi r \qquad$ (see 46, 47, §8·4)

52. $\qquad t Y_0(at) = (2p/\pi r^2)[1 - (p/r) \log (R/a)]$

53. $\qquad Y_\nu(at) = p[(a/R)^\nu \cos \nu\pi - (R/a)^\nu]/r \sin \nu\pi$

$$-1 < R(\nu) < 1$$

54. $\qquad I_0(at) = p/s \qquad\qquad$ (see 12b, §8·1)

55. $\qquad I_\nu(at) = a^\nu p/sS^\nu \qquad\qquad R(\nu) > -1$

56. $\qquad t^{\frac12\nu} I_\nu(2at^{1/2}) = a^\nu e^{a^2/p}/p^\nu \qquad R(\nu) > -1$

57. $\qquad t^\mu I_\nu(2t^{1/2}) = \dfrac{\Gamma(\mu + \frac12\nu + 1)}{\Gamma(\nu + 1)p^{\mu + \frac12\nu}} {}_1F_1(\mu + \frac12\nu + 1;\ \nu + 1;\ -1/p);$

$$R(\mu + \tfrac12\nu) > -1$$

58. $\qquad I_0(ay) = pe^{-bs}/s \qquad\qquad t > b$

59. $\qquad ab I_1(ay)/y = p(e^{-bs} - e^{-bp}) \qquad t > b$

60. $\qquad ab \displaystyle\int_b^t \dfrac{I_1(ay)\,dt}{y} = e^{-bs} - e^{-bp} \qquad t > b$

61. $\quad e^{-bc} H(t-b) + ab \displaystyle\int_b^t e^{-ct} \dfrac{I_1(ay)}{y}\,dt$
$$= e^{-b[(p+c)^2 - a^2]^{1/2}} \qquad t > b$$

62. $\qquad K_0(at) = (p/s) \log (S/a)$

63. $\qquad K_\nu(at) = \pi p[(S/a)^\nu - (a/S)^\nu]/2s \sin \nu\pi$

$$-1 < R(\nu) < 1$$

64. $\qquad t K_0(at) = (p/s^2)[(p/s) \log (S/a) - 1]$

65. $\qquad t K_1(at) = (p/s^2)[(p/a) - (a/s) \log (S/a)]$

66. $\qquad \operatorname{ber} t = p[(p^4 + 1)^{1/2} + p^2]^{1/2}/[2(p^4 + 1)]^{1/2}$

67. $\qquad \operatorname{bei} t = p[(p^4 + 1)^{1/2} - p^2]^{1/2}/[2(p^4 + 1)]^{1/2}$

68. $\operatorname{ber}_\nu^2 2t^{1/2} + \operatorname{bei}_\nu^2 2t^{1/2} = I_\nu(2/p) \qquad\qquad R(\nu) > -1$

69. $\qquad \mathsf{H}_0(at) = (2p/\pi r) \log \left(\dfrac{a+r}{p}\right) \qquad$ (see 12e, §8·1)

70. $\quad i^{-1}\mathsf{H}_0(iat) = \mathsf{L}_0(at) = (2p/\pi s) \sin^{-1}(a/p)$

71. $\qquad He_n(t) = (n!/p^n) \displaystyle\sum_{s=0}^m (-1)^s (\tfrac12 p^2)^s/s!$

$$m = \tfrac12 n,\ n \text{ even};\quad m = \tfrac12(n-1),\ n \text{ odd}$$

72. $t^{-\frac12(n+1)} e^{-a^2/4t} He_n(a/\sqrt{2t}) = 2^{\frac12 n} \pi^{1/2} p^{\frac12(n+1)} e^{-ap^{1/2}}$

73. $\qquad He_{2n+1}(t^{1/2}) = (2n+1)!\,(\pi/p)^{1/2}(1/p - 1)^n/2^{n+1/2} n!$

74. $\qquad L_n(t) = (1 - 1/p)^n$

75. $\qquad e^{-t}L_n(t) = p^{n+1}/(p+1)^{n+1}$

76. $\qquad t^{\nu}L_n^{(\nu)}(t) = \Gamma(n+\nu+1)(1-1/p)^n/n!\,p^{\nu} \qquad R(\nu) > -1$

$\qquad n \geqslant 0$ in (65)–(70)

77. $\qquad \sum_{n=0}^{\infty} \dfrac{2^n n!}{(2n+1)!} D_{2n+1}(2t^{1/2}) = \pi^{1/2}/2\,(p+1)^{1/2}$

78. $\qquad t^{-1/2} \sum_{n=0}^{\infty} \dfrac{2^n n!}{(2n)!} D_{2n}(2t^{1/2}) = \tfrac{1}{2}\pi^{1/2}(p+1)^{1/2}$

79. $\qquad e^{\frac{1}{2}t} D_{2n+1}(2t)^{1/2} = (\pi/2)^{1/2}(2n+1)!\,(1/p-1)^n/2^n n!\,p^{1/2}$

$\qquad n \geqslant 0$

80. $\quad e^{1+\frac{1}{2}t}t^{-1/2} \sum_{n=0}^{\infty} \dfrac{2^{2n}}{(2n)!} D_{2n}(2t)^{1/2} = e^{1/p}(\pi p)^{1/2}$

For additional L.T. see references [13, 14].

REFERENCES

This is merely a brief selection. An extensive list is given in reference 12. Throughout the text, references are indicated by [ ].

1. BATEMAN, H., ' Solution of system of differential equations in theory of radio-active transformation ', *Proc. Camb. Phil. Soc.*, **15**, 423, 1910.

2. BICKLEY, W. G., ' Effect of free surface on compressional shock waves ', *Proc. Roy. Soc.*, A, **180**, 209, 1942.

3. BROMWICH, T. J. I'A., *Theory of Infinite Series*, second edition revised by T. M. MacRobert (1926).

4. CARSLAW, H. S., *Theory of Fourier's Series and Integrals*, third edition (1930).

5. COPSON, E. T., * ' Operational calculus and evaluation of Kapteyn integrals ', *Proc. Lond. Math. Soc.*, **33**, 145, 1930.

6. COPSON, E. T., *Theory of Functions of a Complex Variable* (1944).

7. FERRAR, W. L., *A Text Book of Convergence* (1938).

8. HARDY, G. H., *A Course of Pure Mathematics*, seventh edition (1938).

9. LERCH, M., *Acta Mathematica*, **27**, 339, 1903.

10. LOWRY, H. V., *'Operational Calculus '; *Phil. Mag.*, **13**, 1033, 1932.

11. McLACHLAN, N. W., *Bessel Functions for Engineers*, 2nd ed. (1955).

12. McLACHLAN, N. W., *Complex Integration and Laplace Transform Calculus* (1962).

13. McLACHLAN, N. W., and HUMBERT, P., *Formulaire pour le Calcul Symbolique*, Mémorial des Sc. Math., Fascicule No. 100 (1941).

14. McLACHLAN, N. W., HUMBERT, P., and POLI, L., *Supplément au Formulaire pour le Calcul Symbolique*, Fascicule No. 113 (1950).

15. OSGOOD, W. F., *Functions of Real Variables* (1947).

16. TITCHMARSH, E. C., *Theory of Functions* (1952).

17. ERDÉLYI, A., MAGNUS, W., and OBERHETTINGER, F., *Tables of Integral Transforms*, 2 vols. (1954).

* Signifies that the procedure used in deriving L.T.S. is given.

INDEX